EUCLID'S HERITAGE:
IS SPACE THREE-DIMENSIONAL?

THE UNIVERSITY OF WESTERN ONTARIO SERIES IN PHILOSOPHY OF SCIENCE

A SERIES OF BOOKS
IN PHILOSOPHY OF SCIENCE, METHODOLOGY, EPISTEMOLOGY,
LOGIC, HISTORY OF SCIENCE, AND RELATED FIELDS

VOLUME 52

PETER JANICH
Lehrstuhl 1 für Philosophie,
Philipps-Universität Marburg, Germany

EUCLID'S HERITAGE: IS SPACE THREE-DIMENSIONAL?

KLUWER ACADEMIC PUBLISHERS
DORDRECHT / BOSTON / LONDON

Library of Congress Cataloging-in-Publication Data

Janich, Peter.
[Euklids Erbe. English]
Euclid's heritage : is space threedimensional? / Peter Janich.
p. cm. -- (The University of Western Ontario series in philosophy of science ; v. 52)
Includes bibliographical references and indexes.
ISBN 0-7923-2025-5 (alk. paper)
1. Geometry, Solid. I. Title. II. Series.
QA491.J3613 1992
516.23--dc20 92-34895

ISBN 0-7923-2025-5

Published by Kluwer Academic Publishers,
P.O. Box 17, 3300 AA Dordrecht, The Netherlands.

Kluwer Academic Publishers incorporates
the publishing programmes of
D. Reidel, Martinus Nijhoff, Dr W. Junk and MTP Press.

Sold and distributed in the U.S.A. and Canada
by Kluwer Academic Publishers,
101 Philip Drive, Norwell, MA 02061, U.S.A.

In all other countries, sold and distributed
by Kluwer Academic Publishers Group,
P.O. Box 322, 3300 AH Dordrecht, The Netherlands.

Printed on acid-free paper

English translation of: Peter Janich, Euklids Erbe. Ist der Raum dreidimensional?
First published by C. H. Beck Verlag München, 1989
Translation by David Zook

Printed in the Netherlands

EDITOR'S PREFACE

We live in a space, we get about in it. We also quantify it, we think of it as having dimensions. Ever since Euclid's ancient geometry, we have thought of bodies occupying parts of this space (including our own bodies), the space of our practical orientations (our 'moving-abouts'), as having three dimensions. Bodies have volume specified by measures of length, breadth and height. But how do we know that the space we live in has just these three dimensions? It is theoretically possible that some spaces might exist that are not correctly described by Euclidean geometry. After all, there are the non-Euclidian geometries, descriptions of spaces not conforming to the axioms and theorems of Euclid's geometry.

As one might expect, there is a history of philosophers' attempts to 'prove' that space is three-dimensional. The present volume surveys these attempts from Aristotle, through Leibniz and Kant, to more recent philosophy. As you will learn, the historical theories are rife with terminology, with *language*, already tainted by the assumed, but by no means obvious, clarity of terms like 'dimension', 'line', 'point' and others. Prior to that language there are *actions*, ways of getting around in the world, building things, being interested in things, in the more specific case of dimensionality, *cutting* things. It is to these actions that we must eventually appeal if we are to understand how science is grounded.

What I have just outlined briefly encapsulates the approach of Professor Peter Janich in this book. Professor Janich operates within a philosophical orientation known as *constructivism.* Originated by Hugo Dingler (1881–1954), the position was refined and further developed by the 'Erlangen School' founded by Paul Lorenzen and Wilhelm Kamlah.[1]. Professor Janich was a student at Erlangen in 1961–67. The constructivist looks for presuppositions underlying our employment of the sometimes exotic abstract terms of the sciences, seeking justification for the rationality of science by reducing these terms to presupposed forms of action. One such form is the manufacture of instruments, the 'doing' of science in laboratory and factory. The rationality of science can only be grounded in the possibility of its realization through the performance of experiments; theories are contentless if they cannot be tested. Testing, of course, requires action and manufacture of instruments.

At a deeper level, there are human actions, orientations, interests that are presupposed in developing any rationally viable scientific enterprise. Such activity in the world in which we live is presupposed by science, but is itself presuppositionless. There can be nothing more basic, so the constructivist believes, than fundamental forms of human action. The task of the philosopher is to provide a methodically clear grounding of all scientific terminology in human actions. This book is an example of how the project works.

Dr. Janich has taught philosophy at Universität Konstanz, and served as Visiting Lecturer at the University of Texas (Austin). In 1987 he was a Fellow at the Center for Philosophy of Science, University of Pittsburgh. In 1980 he was appointed Professor of Philosophy in the Philipps-Universität Marburg. The author of several books and numerous papers, his latest work is the book, *Grenzen der Naturwissenschaft: Erkennen als Handlung*, München: Verlag C. H. Beck (1992). The present volume is a translation of his *Euklids Erbe: Ist der Raum dreidimensional?* München: Verlag C. H. Beck (1989).

NOTE

[1] For a more detailed discussion of constructivism, see the introduction to Robert E. Butts and James Robert Brown (eds.), *Constructivism and Science: Essays in Recent German Philosophy*, The University of Western Ontario Series in Philosophy of Science, Vol. 44. Dordrecht: Kluwer (1989), and for samples of constructivist philosophizing, see the papers in the same volume.

ROBERT E. BUTTS

TABLE OF CONTENTS

PREFACE

There seem to be a large number of statements which are distinguished by a peculiar disparity. The greater the intuitive and spontaneous certainty with which the layman, as well as the scientific expert, agrees with such statements, the more unclear the reasons for this agreement become. One could, for example, hardly find anyone who would dispute the following sentence: "A body can very well be at the same place at two different times, but not at two different places at the same time." But what authority is responsible for the undivided agreement over this sentence? Even the philosophical layman can convince himself without much trouble that neither everyday experience nor any branch of science can serve as the foundation. Nor could one logically deduce this sentence from mere definitions of body, place and time, because no alternative definitions and, therefore, no playing room for terminological choices can be found.

It is no different with the statement that empirical space has three dimensions. We know, for example, that any given cupboard, viewed as a body - or as a hollow body - has length, breadth, and height; only one or two of these specifications do not suffice to describe its size or form (or even clearly its position within a room). On the other hand, we know of no fourth determinant of the same kind as length, breadth, and height, but different from these in the same sense as these three extensions are different from one another. Technicians and scientists also use the three-dimensional description - from machine construction and architecture to astronomy - with the same intuitive certainty as the layman.

Properly seen, it is scandalous that in philosophy, particularly for those statements with the greatest consensus, there is the least clarity concerning the reasons for this. Yet it is certainly not the case that philosophers have ignored the problem of three-dimensionality. On the contrary, the list of those who stated an opinion, if often only casually, reads like the table of contents of a general textbook on the history of philosophy. But (in an admittedly fascinating way) these philosophical attempts remain largely fruitless. Not a few are unclear or contain logical errors, while some grasp to wild speculation. The most thorough, and until now most clearly conducted, restrict themselves to reflections on individual branches of science - in particular on geometry and the natural sciences. However none is adequate to explain the obvious success with which the three-dimensionality of empirical space is adopted in technical practice. This technical element was simply forgotten by the philosophers, although the fact that our houses and cupboards have length, breadth and height can be taken neither from transcendental philosophy nor phenomenology, from natural laws nor from the construction of our sense organs.

Criticism for this distressing lack of proof is not to be made against the sciences. Mathematics, with its subsidiary disciplines of geometry and topology, does not treat objects of experience according to its modern self-appraisal. Rather, it treats "space" abstractly, as a set of points. It is therefore disqualified by virtue of its subject matter and methodological limitations. In addition, it can show that it has, based upon its self-defined task, a satisfactory definition for the term "dimension", which nevertheless intentionally avoids a distinction of the triad.

Physics is disqualified because it builds upon a use of instruments for its experiments for which the question of three-dimensionality is already decided from the beginning: instruments are - independent of all further descriptions - always spatially formed bodies, which are described three-dimensionally. Not meant here is the metaphoric use of language, in which every perimeter measured in physics can be called a "dimension" which is a basic (or primitive) term within the frame of a theory. What is meant is the elementary and fundamental understanding of dimension that the physicist refers to when he speaks to technicians and craftsmen about the construction of his instruments.

Other empirical sciences are obviously also disqualified: perceptual or thought psychology as well as biology or brain research. All of these disciplines have always had as their foundation a three-dimensional, Cartesian system of coordinates, or at least three orthogonally paired planes, as the frame for directional orientation.

So it remains a scandal that philosophy has not satisfactorily answered this question until now.

This book makes an attempt to help eliminate this deplorable state of affairs. The reader must judge whether or not this has been accomplished. The goal of this examination, in any case, is to determine what kind of knowledge it is to assume the three-dimensionality of empirical space. It is less a matter of the foundation of a science like geometry or physics than an inquiry into what causes this great intuitive certainty, in spite of the unclarified grounds, on a pre-scientific level.

The hope of saying something new in this field is primarily based on an analysis of historical attempts. Once one has seen that the proper systematic way to understand three-dimensionality results from our daily practice in dealing with artificially formed bodies (that is, "form theoretical"), then the historical attempts show themselves to be trapped in certain prejudices, which have been anchored in science and philosophy since the beginning of ancient geometry. In the end, the greatest obstacle for the previous, failed arguments for three-dimensionality has always proven to be the old problem of philosophy of science: in what relationship does ideal, mathematical geometry stand to the geometrically-described objects of experience?

This book shall therefore bring forth not only a new thesis to three-dimensionality, but based on the example of this problem, raise the epistemological question about the nature of geometry and indicate how validity arguments for deep-seated convictions have at least as much to do with our (language-free) operations as with our speech or thought.

The fact that this book (whose first considerations go back to a lecture in Austin, Texas in 1969) could finally be written, is thanks to the favorable working conditions that I enjoyed during a research sabbatical at Pittsburgh University's Center for Philosophy of Science (Winter Semester 87/88). I am indebted to the former Director of the Center, Nicholas Rescher, as well as the members and guests of the center, to whom I am grateful for stimulating dialogue.

I also wish to express my gratitude to G. Hövelmann, who compiled the literature, as well as the name and subject indexes, and to Mrs. S. Weber, who typed the final manuscript.

Marburg, December, 1988.

PART ONE

The History of the Problem

Introduction

While we have no precise information about this, it would seem that the discussion of three-dimensionality goes back to the beginning of Greek philosophy and geometry. But at the point where the problem appeared for the first time, fully focused in its modern form, the most important theoretical decisions must already have been made. The Greek geometricians discovery that there are, indeed, an infinite number of regular polygons in a plane, but that five - and only five - regular polyhedra are spatially possible, was already available when Plato used the five regular polyhedra (later called "Platonic bodies" for this reason) for an explanation of the perceivable bodies of the physical world in his dialogue "Timaeus."

The ancient proof of the polyhedra statement (in Book XIII of Euclid's *Elements*) is impressive not only for its brilliance but also because it shows which hidden assumptions were already current. The proof makes use of the fact that regular polyhedra are made up of regular polygons, and that all surfaces, edges and angles are undifferentiated. The enumeration of these spatial elements - planes, lines and angles - was thereby considered to be complete. In other words, the geometricians had already accepted the triad of these geometrical objects as an unquestionable assumption.

The venerable age of the problem, and the fact that this problem finds its historical beginning in geometry - which, for its part, has certainly witnessed further development in its two-and-a-half thousand year history - might suggest that there could be an analogous history of the dimension problem. This supposition, however, proves to be incorrect when one examines historical sources. Up to the nineteenth century there were no texts (with the exception of a lost text of Ptolemy) that were dedicated specifically and exclusively to the problem of three-dimensionality. The problem is addressed more often than not during examinations of other, widely varied themes, and then rather peripherally. This peripheral treatment may be the reason that, until recently, no discussion of the dimension problem was conducted as a dispute with (or an improvement of) the older approaches, but always consisted of new *ad hoc* solutions, so to speak. The most recent texts which, in contrast, dedicate themselves to the history of the problem, mostly draw the resigned conclusion that they should not risk making their own systematic propositions at all.

It would, therefore, not make sense to write a history of the problem of three-dimensionality which proceeds chronologically and documents progress in knowledge, or at least changes in understanding. Therefore another way has been chosen: previously proposed answers to the question about the dimension concept and, specifically, the triad of empirical space can be classified according to the fields in which a solution is sought. By comparing the authors of different times, organized in such categories, it is easier to judge whether or not there has been any progress.

We will classify the historically proposed solutions according to the following five types:

1. the purely spatial approaches,
2. approaches which make reference to motion or direction,
3. explanations of three-dimensionality from the laws of force,
4. tracing three-dimensionality back to causality or to more general analytical arguments,
5. perception-theoretical, physiological or biological attempts to provide a basic explanation for three-dimensionality.

Approaches of widely varying directions are included within these categories as well, from strict empiricism to the wildest numerical speculations, from carefully weighed out conventionalism to epistemological naivety. Therefore, this division into five types should not be considered an iron-clad formulation which is responsible for the respective definitions of the concept of dimension, or upon which the triad is based.

A second preliminary remark about the history of the dimension problem deals with the fact that "dimension" is a technical term of mathematics, a modern science. It would therefore be tempting to write the "history of the problem" as the development of the concept which is acknowledged in mathematics today. How - and to what purpose - did topology and analytical geometry come to the definition of dimension which forms the basic knowledge of mathematicians today?

We will not, however, examine this closely related "problem history." It is easy enough to discern that it has nothing to do with the question of this book, which treats the triad of the dimensions of everyday, physical-empirical space. The definition of the dimension number in analytical geometry (as the maximum number of linearly independent base vectors of a vector space), as well as the various topological definitions, is aimed at formulating a concept of dimension, which is *not bound to the triad*, but which makes mathematics in spaces of higher dimensions possible. Everything that would result in a limitation to the dimension number three is intentionally avoided in mathematics.

Epistemologically, this process of mathematics can be understood as a conscious limitation to the criterion of logical compatibility. There are some very good reasons for this. It does not lead to a logical contradiction, for instance, to practice a five-dimensional analytical geometry. Every point is labelled with five coordinates and, as far as this is at all desirable, the demand for mutual orthogonality of coordinate axes can be fulfilled with a corresponding criterion for orthogonality. The fact that one cannot intuitively imagine five lines, all paired at right angles to each other, is no argument against this geometry. For it is not its goal to provide a theory for our intuitive concepts or for our technological experiences and inventions (as in dealing with straight rods which lie at right angles to each other).

The consequences of the formalistic self-conception of modern mathematics will be addressed later in this book, so they need only be mentioned here. This self-conception affirms that modern mathematics is a science which deals with symbols (formulas) of its own creation, and therefore makes no claim to copy structures of reality. Correspondingly, mathematical disciplines such as geometry and topology (in which, for historical reasons, intuitively understandable words appear) also lose, quite intentionally, their direct reference to reality. "Spaces" are only sets of points of differently defined structures, and have at most in a few special cases (now only historically understandable, and of unimportant content) reference to spatial perception or experience. Even the fact that one of the founders of topology, Henri Poincaré, saw three-dimensionality as the most important question for the new discipline and as motif to its development (instead of the designation "topology," Poincaré used the expression "*analysis situs*," which goes back to Leibniz) is now only a footnote to the history of science, without systematic significance. We will therefore not deal with the history of topology at all (in which, apart from Poincaré, yet another significant author of the philosophy of science must be mentioned, namely Luitzen E. J. Brouwer).

The disqualification of modern mathematics from the problem of three-dimensionality for reasons of its own self-chosen principles, finds a parallel in the disqualification of modern physics: here too an unbiased layman, uninitiated in the subtleties of the higher methodological self-conception of physics, might expect a solution. But according to its modern self-conception physics is an empirical science which continually holds even its own conceptual foundation open to empirical revision. Thus, the transition from the classical physics of Newton to Einstein's relativity physics is normally understood as a revision of the fundamental terms of space, time, causality, etc. through observation and experiment, i.e., through physical experience.

But in the history of physics there is neither a *foundation* for the fact that physicists always conceive a body, its position and motion, with three spatial dimensions, *nor* was there ever any doubt about its validity, or even a revision of the concept (with the exception of a discussion from Galileo, which will be taken up later). Physics is essentially naive with regard to the three spatial dimensions: it does not make this an object of one of its (empirical) theories, nor is its assumption subjected to a critique. Physics only engages in meta-observations in which the logical compatibility of the assumption of three spatial dimensions with other, already acknowledged, physical theories is verified. And so it was when Paul Ehrenfest, in a lecture at the convention of the Academy of Sciences of Amsterdam in 1917, proposed the thesis that the stability of closed planetary orbits in a gravitational field in accordance with Newtonian theory is only valid for three spatial dimensions (and analogous for other orbital motions in similarly structured fields). We will go into this in Chapter Three.

Physics has abstained from any epistemological critique or methodological clarification of its own practice; for instance, when characterizing the position of bodies in a laboratory or astronomical context through three dimensions. The consideration of consistency in analytical science (following Ehrenfest's example - that is, to question just how far three-dimensionality is logically compatible with acknowledged theories) is essentially not much further developed than Kant's concept (to be discussed later).[1] This states that three-dimensionality is exhibited in the *form of the laws of force*. (All force which is given off by a punctiform abstract body, must propagate through a sphere around this body. The surface of this sphere grows with the squared radius, corresponding to the relation of the gravitational attraction between two bodies.)

With this, the reasons have been listed as to why the methodical self-conceptions of modern mathematics as well as modern physics prohibit one from expecting a solution to the three-dimensional problem from either of these sciences. Therefore, the history of mathematics and physics is obviously not the same as the history of the questions and proposed solutions concerning the three-dimensionality of empirical space. In short, this matter is not a mathematical problem, and physics has always supposed the problem to be solved, or else has never made this question into a problem in the first place.

---

1 Cf. Part I, Chapter Three, p. 59

CHAPTER ONE

The Purely Spatial Approaches

WHAT DOES "PURELY SPATIAL" MEAN?

Of course all considerations of the dimension problem are "spatial," or at least they never manage without spatial metaphors. However, the categories 2 to 5 listed in the introduction bring many additional definitional or descriptive means to bear. Therefore, the approaches which are treated in this first section have been named "purely spatial." In these neither motion nor force, propagation nor conditions of perceptions are considered. This is at first, of course, a negative distinction.

As long as one does not hold any particular systematic scruples, one could understand "purely spatial" in a positive sense as simply meaning "geometrical." Accordingly, we would have to reckon words like surface, line, and point just as much to the purely spatial expressions as length, angle, equally long, congruent, plane, straight, etc. There is, however, no unanimous opinion as to just what "geometry" is (and this will not surprise only laymen). According to the view of modern mathematicians, geometry is a mathematical subtheory like others, such as algebra. As such it is a strictly formal field, which is to say, a field which manages without "concrete" definitions of its basic concepts. The fact that the mathematical layman might associate the surface of a good mirror or the edge of a good ruler with words like plane or straight has nothing to do with the geometry of the contemporary mathematician. This association is simply a historical accident.

However, for the modern physicist, geometry consists of every concrete interpretation of the formula system supplied by the mathematicians. A simplified example of this would be the interpretation that light rays are regarded as straight lines, and that points can be realized in small diaphragms or crosshairs, but also in far less evident interpretations. Whether theorems of geometry are valid becomes a question of scientific experience, according to the contemporary view in physics. Then again, both views - the "formalistic" view of the mathematicians, and the "empirical" view of the physicists - are relatively young, and they encounter logical, methodical, or other reservations in some philosophers. This book shares such reservations, and favors a more thoroughly secured, methodical understanding of geometry, as presented in Chapter Six and Chapter Nine.

Still far more difficult, because it is also tied to problems of interpreting ancient texts, is the question concerning which understanding of geometry was present at the beginning of the history of Occidental thought, especially in the case of Eudoxos, Euclid and *the* philosophers, Plato and Aristotle. There is, in fact, a rich literature on the history of philosophy about this, but in regard to our dimension problem it is of little help. The reason for this lies, above all, in the fact that most historians are not aware of the systematic problems involved in founding a geometry and consequently do not approach the old texts with questions about these problems.

In our context the decisive question is, foremost, about the relationship of the geometrical words, as they are found in Euclid's writings, to real bodies. It so happens that the Greek words for body (*soma* and *stereon*) are really not strictly differentiated terminologically. A cube can just as well be *soma* or *stereon*, and with these terms it is in no way certain that one can express this difference; that the former refers to something like a cube-shaped stone, and the latter to an imagined, mathematical cube, i.e., a geometrical figure.

Therefore, at the beginning of our discussion of "purely spatial" determinants, we want to allow both the *geometrical* vocabulary (whereby it is always necessary to specify the geometer or geometry book - and hence the understanding of geometry - to which the reference is made) as well as a vocabulary for the spatial properties of *real bodies*. With these we are accustomed to distinguish surfaces, edges, angles and further forms. Indeed, this was already as natural for the speaker of ordinary language in ancient Greece as it is for us today.

A further problem raises the question of whether "motions" are to be classed under purely spatial determinants or not. "Real body motions," like that of a thrown stone, shall not be considered here; they will be discussed in Chapter Two. But what about the "motion" of imagined, mathematical, or ideal forms? It is well-known that there are pairs of triangles, whose corresponding side lengths equal each other completely, which, however, cannot be superimposed by shifting them in the drawing plane. They are not congruent, but only "mirror-congruent." Intuitively speaking, one would have to cut one of the triangles out of the paper, lift it out of the drawing plane and turn it over, in order to be able to lay it over the other, covering it. (Even early on, Euclid was unable to manage without the use of such "flips," which were then later seen as motion through a "higher dimension." But he certainly tried to forgo this means of construction, or proof, for as long as possible.) We will also address instances of "ideal motions," together with the motion of real bodies, although such flip motions are replaced by a definition of mirroring, and therefore can be worked around so that no more motion actually occurs. Thus, only the statical considerations shall be classed under the "purely spatial" type, be they formulated in the "ideal" language of geometry or in the language of the craftsman and technician for real bodies.

THE DISTINCTION OF THE NUMBER THREE

Aristotle is generally credited with being the first to recognize the problem of three-dimensionality as such. Often claimed as the founder of almost all the modern sciences - from biology to psychology, literary studies to political science, logic and linguistics to cybernetics - Aristotle was, as far as is known, the first to seize the triad of bodily extensions as a topic, in the introduction to his text "On the Heavens."

Of course, the fact that this achievement is attributed to Aristotle must be credited more to the general historical significance of Aristotle, and to his other philosophical achievements; or perhaps to the fact that Galileo later responded critically (scornfully, in fact) to Aristotle's contribution.

In the first place, the Aristotelian treatment of the dimension problem took up only a few sentences - less than a modern book page - which appear rather incidentally in the introductory remarks to the text. Secondly, they are in many respects unclear, and by no means convincing. And thirdly, it is not clear that any original contributions of Aristotle's can be found in these sentences at all, much less any of importance.

The text "On the Heavens" begins with the indication that natural science is concerned with bodies and magnitudes, as well as their changes and the principles of these changes. Then Aristotle refers to a truly significant part of his natural philosophy, namely his continuum theory, which he developed in "Physics."[2] He reiterates that the continuum is defined as "that which is divisible into parts always capable of subdivision." He continues that the singly divisible magnitude is the line, the doubly divisible magnitude is the surface, and the triply divisible magnitude is the body (*soma*). This is followed by his first "argument" for the distinction of the triad, with reference to the tenet of the Pythagoreans, which states that the whole world is determined by the triad, since everything has a beginning, middle, and end. In other words, he calls upon an outside authority.

His second argument is strikingly modern in contrast: Aristotle observes ordinary language and discovers that we have a separate word for combining two items (in English "both," in Greek "*amphi*"),

2 A short presentation of the Aristotelian continuum theory can be found in Janich (1980), pp. 250-254

while we are not accustomed to saying "all two." In the case of three items, however, we do say "all three," so that three represents a completeness or perfection. (The trivial fact that we can just as easily say "all four" is not taken into account.)

After the triad appears to be distinguished in this way, the body is called the only *complete magnitude* for having brought this triad to completeness in the form of three extensions. Here the Aristotelian doctrine of principles comes into play, because all change in nature has to occur in accordance with teleological principles, among others. It corresponds to a natural law when the body achieves its purpose through its triple extension.

It is important to see that Aristotle *begins* his argumentation with the distinction of the triad, and that he binds his analysis of the three extensions of bodies (as opposed to surfaces and lines) to the supposition that this distinction was achieved. Particularly significant is that he does not do it the other way around, that is, first examine the divisibility of bodies in order to somehow arrive at the triad. Much rather, the triple extension of bodies simply confirms anew a principle, according to which the triad may already be seen as distinct.

Therefore, the thought - by no means restricted to Aristotle - that the triad is in itself distinguished in a special way will be the focus here. The problem of the division or divisibility of bodies, and the question of how physical things are set in relation to ideal objects like surfaces or lines, will be tabled until later.

Galileo, in his "Dialogue Concerning the Two World Systems,"[3] allows Salviati, the representative of his own view, to respond to Aristotle's argumentation:

> To tell you the truth, I do not feel impelled by all these reasons to grant more than this: that whatever has a beginning, middle and end may and ought to be called perfect. I feel no compulsion to grant that the number three is a perfect number, nor that it has a faculty of conferring perfection upon its possessors. I do not understand, let alone believe, that with respect to legs, for example, the number three is more perfect than four or two; neither do I conceive the number four to be any imperfection in the elements, nor that they would be more perfect if they were three. Therefore it would have been better for him to leave these subtleties to the rhetoricians, and to prove his point by rigorous demonstrations such as are suitable to make in the demonstrative sciences.

---

3 G. Galilei (1956), p. 11

Seen historically Galileo's critique, his mockery, was of no great use. The distinction of the three-dimensionality of space based upon a preceding distinction of the triad has been continuously repeated.

As Adolf Trendelenburg[4] writes in his "Logical Inquiries" of 1862, Hegel, as opposed to Kant, had adopted the following argument:

> The necessity of the three dimensions is based upon the nature of the concept, the determinations of which, however, to the extent that they present themselves in the abstract quantity in this first form of the examination, are a merely superficial and completely empty distinction. Therefore, one could not say how height, length, and breadth are distinguished from one another, because they are only *supposed* to be distinguished, but *are* not distinctions.

After this quotation from Hegel, Trendelenburg continues:

> The three characteristics of a concept are generality, particularity, and uniqueness. True, the number of dimensions corresponds to the number of these characteristics. It would have been necessary to show further, however, how these three characteristics of a concept, have, each for itself, extrinsically produced the three dimensions. ... It would have been necessary to show how generality specifically corresponded to one dimension, particularity to another, and uniqueness to the third... As it is, hardly a vague analogy is left here.

The Aristotelian and Hegelian examples, with the respective criticisms of Galileo and Trendelenburg, should suffice to indicate that attempts to show a special quality of the number three, detached from the number of spatial dimensions (in order to prove the latter), belong to the realm of crude numerical speculation. They cannot be proven, and hardly seriously discussed, because there are no proposals available as to how the leap from a distinctive number to the properties of space or bodies should look in argumentative form. Especially hard to see, is how such an attempt should gain credibility for our everyday understanding of three-dimensionality or for technical-physical practice. We will therefore turn to another attempt which is based more closely upon ancient geometry. Here again, Aristotle is the most important author.

---

4 A. Trendelenburg (1862), pp. 227-228. (Unless otherwise indicated, the English translations for quotations from German works are by the translator of this volume. The page numbers are from the original texts.)

DIVISIBILITY AND DEFINABILITY: ARISTOTLE

More important and productive than the aforementioned arguments for the distinction of the triad, is Aristotle's discussion of the *division of bodies* and, within that context, of *surfaces and lines*. We will, of course, have to draw on other Aristotelian texts because the introduction to "On the Heavens" is unclear about how the relation between physical and geometrical-ideal entities is intended. For body, Aristotle uses the word "*soma*," which is applied as a predicate for stones and pieces of wood, as well as for the bodies of living beings. Therefore, "to divide" is an action, like that of a craftsman, for example, the cutting apart, or sawing to bits, of real things. What is striking here, is that Aristotle does not hold to the order of steps that every person who conducts a real division must go through. Obviously, the action of division must always begin with the division of a body in order to proceed from there to surfaces and then, perhaps, to lines. Aristotle, in contrast, mentions first the line, as singly divisible, then the surface, as doubly divisible, and finally the body, as triply divisible. He explains neither here, nor elsewhere, what single, double and triple mean. It also remains unclear what the Greek words for surface (*epipedon*) and line (*gramme*) are supposed to mean (which are, incidentally, exactly the same words that Euclid uses).

One must assume that the Aristotelian terminology in the words "to divide," "body," "surface" and "line" was either supposed to cover both speech about physical things and about geometrical entities, or that no such difference was intended.

Since it is largely unclear what Aristotle is thinking about when he speaks of division, it would be appropriate to take a look at his discussion of the continuum, where he successfully comes to terms with the position held by Zeno of Elea and demonstrates mistakes in his famous paradox. Unfortunately, this only makes the situation even more confused. His argument here (of itself completely persuasive) is that something divisible (like a line segment) always produces when divided, again, something divisible. Otherwise an extensive, divisible magnitude could be made up of indivisibles (as Zeno assumed in the paradox of the flying arrow, which stands still at every point in its path). Aristotle can show, with support of the then current definition of point ("a point is that which has no parts"), that points can only lie next to each other in such a way that they either fall together into one and the same point, or they lie "separate," which means that another point can always lie between them. This is the main reason why points cannot touch each other like pearls in a pearl necklace to form a line segment.

This Aristotelian argument shows that it is undoubtedly a *geometrical* point spoken of here, i.e., an ideal construct, as every real body is continuous (according to Aristotle), and is thus repeatedly divisible. Therefore, a point clearly cannot be a small body or a spot painted on a body, and can also not be achieved through the continuous division of such bodies or spots.

It follows from this that Aristotle remains inexact. "Dividing" and "divisible" refer to both the division of physical things as well as mathematical-ideal constructs. Whether Aristotle holds that three-dimensionality is a property of physical things, or only of geometrical objects, remains impossible to ascertain.

But, to hold on to a positive result, at least it can be deduced from the Aristotelian texts that the number of the dimensions (for which Aristotle used the word *diastasis*) is determined by the *number of possible divisions.* Beyond that, one can only guess as to how the number of *possibilities* is to be defined. It remains an open question whether Aristotle would have agreed with the view that *cuts* through bodies produce surfaces, that cuts through surfaces produce lines, and that cuts through lines produce points. On the other hand, the conjecture that Aristotle would want to form lines from points, surfaces from lines, and bodies from surfaces, is out of the question because of the Aristotelian argumentation in the continuum theory.

Hippocrates G. Apostle's book on Aristotle's philosophy of mathematics[5] brings some light into this darkness. It is shown there that Aristotle distinguished between two types of priorities regarding division and circumscription, namely "priorities concerning definition" and "priority concerning substance." Points are (among other things) boundaries of lines, lines are boundaries of surfaces and, by analogy, surfaces are boundaries of bodies. Now if one thinks of "division" as the act of cutting, then the corresponding boundaries will be produced through divisions. According to Apostle, it was Aristotle's opinion, regarding *defining*, that the boundary has precedence over the bounded, which means that the point is to be *defined* before the line, the line before the surface, and the surface before the body. Aristotle advocates thereby the definitional sequence as it is actually found in Euclid (Cf. p. 16). The Aristotelian reason for this is that it would not attain the full essence - or a satisfactory definition - of the point, to merely see it as the boundary of a line (according to Aristotle, objects like a point *are* what they are *defined* to be). Points have independent properties as well, e.g., to have no parts or to be indivisible, as in the usual definition. (The thought that the indivisibility of the point results from a definition of the point as the intersection of two lines occurs to neither Aristotle nor Euclid.) By

5 H. G. Apostle (1952), pp. 97-102.

analogy, lines are not merely boundaries of surfaces, and surfaces are not only boundaries of bodies, but are always something else as well.

Regarding *substance*, though, the priorities are reversed: there the bounded has precedence over the boundary. But what does "substance" mean in this context? And in what sense does the bounded "precede" the boundary? (In regard to the definitions, the word "precede" pertained simply to the definitional sequence.) Aristotle's answer leaves one unsatisfied because, in the end, it offers nothing more than this: that one can only speak of circumscription or of boundaries when there is something to be bounded. In other words, for Aristotle, it is only a matter of *linguistic* steps here also, in the sense that every explanation or definition of "boundary" (or of a certain type of boundary, like the "point") must presuppose something (already explained or defined) which is to be bounded.

The linguistic character that even the priority of substance finally exhibits does not lead to a logical incompatibility of the two sequences. Because the priority of substance now means that, as far as the point is to be defined *as the boundary of something*, this something must already be defined beforehand. On the other hand, as stated above, the point should not be primarily defined as a boundary. In any case, we realize "later" that, among other things, we form the boundaries of lines with the (already defined) points.

By differentiating between the two priorities of the boundary and the bounded the hope is that, firstly, their *generation* (e.g., cuts through bodies, as a physical field of application for geometrical distinctions) and, secondly, their *definition* might be found in a deductive geometrical theory. Unfortunately, this hope remains unfulfilled.

Another consideration of Aristotle's seems to bring us further here. He asks whether or not the three entities - bodies, surfaces and lines (now there is no mention of points!) - are of the same kind (or equivalent) by their nature. The subject of Aristotle's "Physics" are those natural things which have the beginning and cause of their motion within themselves. Within this framework, physical *bodies* as the raw material for the dividing process are primary. Surfaces and lines, in contrast, appear *on* such bodies. They do not exist independently. A surface is only divided by a cut when the body on which the surface appears is divided. Thus, according to Aristotle, there is a connection between the division of a physical thing, as in dissection, and the division of ideal objects like surfaces and lines. Only through the division of bodies can, in primary dependence, surfaces also be divided, and, in secondary dependence, lines.

However, this insight is not used in any Aristotelian text to justify a definition-theoretical position for the foundation of geometry. It remains isolated and inconsequential. An operative generation of spatial forms on physical bodies and the definitional construction of a geometry are for Aristotle two entirely different areas; an operative solution of the dimension problem is, therefore, obviously inconceivable for him.

In summary, the writings of Aristotle do not yield a convincing thesis to the problem of three-dimensionality. A proof for the triad of extensions is lacking. The short remarks in the introduction to the texts of "On the Heavens" do not achieve this proof. Furthermore, Aristotle simply repeats that which was already convention among geometricians of his day, that is, one either comes up with *three generators for bodies* (point, line, surface), or with *three objects to be divided or bounded* (body, surface, line), when one does geometry. It remains unclear, when speaking about bounding or division, to what extent this refers to real bodies or to ideal mathematical objects. All that is to be found are a few definition-theoretical considerations, to which, from the modern perspective, it must be critically noted that the proposal to proceed, in definition, in an ascending sequence from point to line to surface is only postulated, never carried out, by Aristotle. Indeed, his continuum theory precludes such a procedure.

At any rate, the sequence proposed by Aristotle was also the one found in Euclid's *Elements*. Therefore, we shall now take a look at ancient geometry from the perspective of the problem of three-dimensionality.

ANCIENT GEOMETRY AND THREE-DIMENSIONALITY

The *Elements* of Euclid begin as *planimetry*, that is, as geometry of (or in) the plane. Stereometry (the geometry of spatial figures) first appears in Book XI. It is there that "spatial" (in the sense of non-planar) forms are first defined. This arrangement, as well as the definitions in Book XI, shows that Euclid proceeds "constructively," in the sense that first the point, then the line and surface are examined as objects, *before* higher-dimensional objects are constructed from them. Properties of non-planar forms are consequently traced back in definition to those of planimetry. This is the case when, in Book XI, Definition 4, the orthogonality of two planes is defined by the orthogonality of two lines. The sphere (Definition 14) is defined by the rotation of a semi-circle around its diameter, and the cone is defined by the rotation of a right triangle around one of its perpendiculars.

In order to be able to better judge the building blocks of these definitions from Book XI, we will first take a look at the definitions of the (planimetric) Book I. Important here are the first five definitions:

1. A point is that which has no part.
2. A line (is) breadthless length.
3. A straight line (is) a line which lies evenly with the points on itself.
4. A surface is that which has length and breadth only.
5. A plane surface is a surface which lies evenly with the straight lines on itself.

Already the ancient commentators on Euclid complained that not even Definition 1 can be a satisfactory definition. It formulates such a restriction, such that one then knows that points are indivisible. However, the statement contains no equivalence of definition, that is, it does not allow for the replacement of the defined word by the definiens, i.e. another, possibly longer expression (with known words). This is also true for the definitions of line and surface.

The *homogeneous* lines and surfaces, that is, the straight lines and planes, are determined through the "lying evenly" of points and lines, respectively. At least here an understandable sequence is followed: in later definitions, use is made of the words defined earlier. On the other hand, it is not further explained what "lying evenly" is supposed to mean. Even the incidence of, respectively, points and lines or lines and surfaces, remains unexplained, and obviously numbers for Euclid among the comprehensible words, which is quite astounding. After all, how should a relation of two items to each other be understandable, when these themselves are held to be in need of definition?

A further critique that was already raised in ancient times, concerns the negative form of the Euclidean definitions. It is stated only what points and (only partially) what lines and surfaces are *not*, or do *not* have. Within our context, there is nevertheless something of interest in this. Namely, that in these negative definitions precisely those unexplained words appear which are partially needed again in Book XI: "part," "length," "breadth," and "depth." The definitions of Book XI begin with: 1. A body (*stereon*) is that which has length, breadth and depth. 2. The boundary of a body is a surface.

Obviously, whatever one might have to say about the logical and definitional perfection of these definitions, for Euclid and the ancient geometricians it was *at least permissible* that the words "part," "boundary," "length," "breadth," and "depth" appeared in the definiens as undefined words, that is, words which were known and usable for purposes of definition. When one looks at the Greek originals of these words, one can assume that they belonged every bit as much to the everyday language as do those of the corresponding English translations today. It would be false to deduce from this that Euclid had an operative understanding of geometry, such that the basic concepts of geometry arose from artisan methods for production of spatial forms. Nevertheless, it is certain that the geometrical terms which are explained in the definitions were at least supposed to become *more comprehensible* by referring back to elementary spatial terms and actions of everyday life. There is much evidence, for example, which would suggest that the generation of spatial figures by rotating plane figures (such as generating a sphere by rotating a circle on its diameter) is taken from the techniques of stonemasonry.

In view of the consequences for the definition of the term "dimension" and for the question of the triad, the consideration of Euclid's "Elements" yields little which goes beyond the opening remark to the beginning of the history of the problem. Where geometry crystallizes - as Plato applies it to his theory of sensory qualities in Timaeus - it is already current as unquestionable knowledge at the everyday level that bodies have length, breadth and depth, that they are bounded by surfaces, edges and corners, and that there is an intuitive, but not explicitly explained, parallel between the geometrical terms "body," "surface," "line," and "point," and the everyday words "body," "surface," "edge," and "corner."

It can already be seen here, that these parallels raise, above all, the following problem. On one hand, *a point is, by definition, not further divisible*. On the other hand, even after a corner has been produced by three cuts or divisions through *a physical body*, that body can, indeed, be further divided (by all means) by a cut which goes through its corner. And so - quite apart from the more demanding considerations of what a definitional process might look like, which establishes a clarified link between real and ideal language - the incompatibility which exists between the divisibility of a body in a corner, and the indivisibility of a point (an idealized corner), must be overcome here.

What cannot be "proven" or otherwise inferred from Euclid's geometry, is a concept of dimension, or a foundation for the triad. The Greek word for dimension is diastasis. This is not one of Euclid's geometrical terms. And there is no consideration, much less a proof, that at most three planes can intersect at right angles to each other, or that no more than three straight lines can stand perpendicular to one another. The important role that the concept of division plays for Euclid as well as for Aristotle, is of absolutely no consequence for this problem.

THE PLACE OF A BODY: FROM ARISTOTLE TO GALILEO

In the Platonic tradition, empty space - as a container - is the prerequisite for the existence of bodies. With this in mind, one could now make a direct inquiry (concerning the dimension problem) into "space" itself, without regard for the divisible bodies found therein.

Can an argument for three-dimensionality be won by inquiring into the place and motion possibility of bodies?

An explicit theory of place is found for the first time in Aristotle's "Physics."[6] Within the framework of his program of research into principles Aristotle was clearly forced to develop such a theory because he had singled out the concept of change as the fundamental concept of his natural philosophy. "Natural" or "by nature" is, for Aristotle, that which carries within itself the beginning or cause of change. Among changes, the changes of place are an important, if not the most important, subgroup, so that a consistent language for place changes had to be prepared. Aristotle solved the problem which arises thereby - that is, that bodies have places and these can change, but that places themselves cannot have places, or change them - with a peculiar definition, according to which the place of a body is the inner boundary of the first unmoved body.

The famous Aristotle commentator, Simplicius (whom Galileo later memorialized in the figure of Simplicius in his *Dialogue Concerning the Two World Systems*) grappled with Aristotle's place theory, and demonstrated that it advances a *two-dimensional* definition: according to Aristotle, the place of a body is a surface. For Simplicius this is, in reference to Aristotle's remark concerning the three-dimensionality of bodies, quite simply an "*aporia*," a contradiction. How can a three-dimensional body have a two-dimensional place?

Depending on the purpose of such a theory of place (like that of overcoming the paradox that the cosmos, as a whole, can be neither moved nor unmoved), it is quite possible to bring both parts of Aristotle's theory to a logical consistency: to wit, the three-dimensionality of the body, and its two-dimensional enclosing surface as its place. This can happen through an appropriate repair of the terminological proposal of the Aristotelian theory. But then, it must be seen that even making such a change would provide no new or additional knowledge about space or the set of places and their dimensions. In short, nothing relevant to the dimension problem could be added to the thesis of the three-dimensionality of bodies from the fact that now the place of the bodies is also spoken of. In other words, there is no particular, additional dimension determinant "of space," or of place or places.

As far as we know, the idea of defining the three-dimensionality of space with the division of, or cuts through, bodies, is first taken up again (although still not successfully) in the nineteenth century. We return therefore, to Galileo's critique of Aristotle which, in fact, was

---

6 Aristotle, Physical Lectures, Book IV

meant as the introduction to Galileo's own proposal; a proposal which is presented in the *Dialogue* by Salviati.[7]

Salviati's argument, threading its way through the dialogue, amounts to this: the straight line is distinguished from all other possible lines (i.e., curved lines), as the distance between two points. By asking a question of him, Salviati has Sagredo (the representative of everyday common sense) give the reason for this. It is because a straight line is the shortest distance between two points, and because every measurement, "ought to depend upon that which is unique and definite." Then they discuss what would best determine the distance between two parallel lines, coming to the conclusion that it would be a line perpendicular to both. In Salviati's words, "So now we have it that the first dimension is determined by a straight line; the second (namely, breadth) by another straight line, and not only straight, but at right angles to that which determines the length. Thus we have defined the two dimensions of a surface; that is, length and breadth." Finally, the height of the room is asked, and the length of a plumb-line is accepted as the answer.

Salviati summarizes these considerations:

> Therefore if you assign any point for the point of origin of your measurements, and from that produce a straight line as the determinant of the first measurement (that is, of the length) it will necessarily follow that the one which is to define the breadth leaves the first at a right angle. That which is to denote the altitude, which is the third dimension, going out from the same point, also forms right angles and not oblique angles with the other two. And thus by three perpendiculars you will have determined the three dimensions... by three unique, definite and shortest lines. *And since clearly no more lines can meet in said point to make right angles with them, and the dimensions must be determined by the only straight lines which make right angles with each other, then the dimensions are no more than three* [My italics. P.J.]; and whatever has the three has all of them, and that which has all of them is divisible in every way, and that which is so, is perfect, etc.

There are a number of aspects to underscore about this proposal:

1. The concept of dimension is explained or defined within the framework of a *geometrical* language. It is *straight lines* which form the dimensions (labelled as length, breadth, and height).
2. The three straight lines which are labelled as length, breadth and height, must stand *pairwise perpendicular to each other.* Oblique coordinate systems are not intended. Orthogonality is likewise

7 Cf. Note 3

geometrically established. The distance between two parallels (the breadth) is the length of the perpendicular between the two. (It is clear that the perpendicular is unequivocal, and the geometry is unmistakably Euclidean.) Apparently, there are *only three* such straight lines.

3. The three-dimensionality of *bodies* is a derived characteristic, and is not primary in definition.
4. Finally, the distinction of the length and the straight form of the lines, as well as their use as the three spatial axes, is achieved with a reference to *measurement*.

Questions of precedence, in which, for the most part, there is already a lot of arbitrary interpretation, are of no concern here. Whether or not Galileo was the first to advocate the said opinion in each of these four points can remain open.

Where historical effect is concerned, however, it is hardly possible to overestimate these four points, even today. This is true even with the inclusion of their highly problematic assumptions. First of all, Galileo inquires neither here nor elsewhere about the availability of the necessary geometrical language. Each partner of the dialogue, every reader, must already know what a straight line, a right angle, etc., is. With that, the thought can no longer arise that the definitional provision of the geometrical language, and the determination of the concept and number of dimensions have something to do with one another. The proof of orthogonality (Point 2) is only implied, and arose from the practice of measurement in craftsmanship. One always takes the shortest line, such as would be formed by a tightened string between two parallel beams or walls, as the "distance" between two points. It does not seem to concern Galileo that this argument only succeeds for breadth or height, when parallel lines are already given in a plane, or in a square-shaped form such as a room. The question as to whether there could not be more than three lines perpendicular to each other, is negated alone by the power of the word "apparently." This fact, then, must be evident to all, and be neither capable of proof, nor in need of it. Finally, the property of bodies of possessing three (and therefore all) dimensions is held to be the same as divisibility in all directions. No further particulars are to be found regarding whether one knows from experience that bodies possess all three dimensions, or what this has to do with the property of divisibility. The most generous reading allows one to take from the text that the triple divisibility of bodies presupposes *three different* directions of division, perpendicular to one another.

So it remains fully open as to which insights or grounds are responsible for the triad of dimensions, apart from the evidence that there can be no more than three straight lines perpendicular to one another. The only argument which remains from the entire set-up is the starting point, in which the distinction of the straight line, with the help of which "dimension" is then defined, is based upon a reference to measurement. The straight connection of two points is the shortest. Galileo is the originator of an unfortunate methodical confusion, which has lasted to this day.

SPATIAL FORM AND MEASURABILITY

Galileo has Sagredo respond to the question concerning which of several lines between two points determines their distance from each other - and why - with the answer that it is the straight line *because* it is the shortest; and that it is "unique, distinct, and determinant." So "straight" is not defined here by the "shortest connection," as determined by some measurement process; rather, it is already *known* from geometry that the straight line is a distinctly defined line (in modern terms, that exactly one straight line is determined by two different points).

In addition to that, however, one must know that straight lines are always the shortest connection for any two of their points. But upon what is this knowledge based? Surely not upon the experience of measurement. For Galileo, like any craftsman, then or today, would not have taken seriously an actual measurement result according to which a crooked connection of two points is shorter than the straight one. In other words, "straight" and "shortest connection" were taken *de facto* for synonymous. The weakness (already spoken of above) of the Euclidean definition of the straight line no longer plays a role for the application of geometry in the measurement of bodies and their motions. Now it can be determined by measurement which connection is the shortest, and therefore "straight." This lends the definition of dimension and the thesis of its triad an empirical turn, whereby it is to be determined by the experience of measurement whether a line (in this case to be understood as drawn) is straight, or whether two straight lines are perpendicular to one another. This conception of Galileo's, even today, forms the core of most modern convictions. It contains, however, a decisive error.

How should the distance between points be judged as longer or shorter? In particular, how can it be decided for two non-coinciding lines with coinciding endpoints which is shorter and which longer? The standard answer, as given in the prevailing empirical doctrine of today (even seeping into physics books in our schools, thus providing for the further spread of this opinion), is stated: Measurement is counting how often a standard unit fits within a given magnitude. But how is this to be concretely verified? There seems to be, in principle, two possibilities. One can either take a drawing compass with a fixed angle, using the distance between its two points as the unit length, and then tick off the entire line, point for point, and segment by segment, counting thereby the number of units. This procedure has the disadvantage that one can only approximate the length of a curved line with straight segments. (If the curved line were formed by a completely inflexible string, then this string would be longer when laid in a straight line than the length found by the compass in its curved state.) Or one uses a straight rod of the unit length and lays it on the line to be compared so that it always stays in contact with one point, but never slides against the line.

In both procedures, however, it must be provided for that the unit of length does not change during transport, i.e., that the compass or ruler is "rigid." This is normally not a problem for the craftsman. He knows which materials and which procedures are appropriate for his task. But the physicist must know more, because he makes many calculations based on measurement results, which lose all meaning if the measurement magnitudes don't have the same logical properties as his figures. For instance, he must know how to check the rigidity or immutability of moved measuring rods - and it is not a reasonable procedure here to use a second measuring device to check the first, because exactly the same problem of transport dependence would apply to it.

In addition, the physicist must know which operations with rulers correspond to arithmetical operations with numerical measurement results. If the equation $1 + 1 = 2$ is to be valid for the addition of lengths, then the rods may only be laid end to end in a straight extension; if one laid them at an angle of 60° to one another, then one would have a situation in which the appropriate equation would be $1 + 1 = 1$. In other words, in order to be able to judge whether a line is straight, one must be able to measure, but in order to be able to measure, one must already be able to judge if a line is straight. This is a circle.

Briefly, it cannot be determined through measurement whether a line is the shortest connection between two points, and is therefore straight (in the sense of the definition); instead, a definition of straight (and a reference to properties of real bodies, e.g., "straight" rods, rulers, etc.) must already be available before measuring rods can be properly produced.

This method-critical digression in reference to the Galilean proposal for the solution to the dimension problem is the second indication - following that of the logical inconsistency within the ancient conception between the indivisibility of a point and the divisibility of the corner of a body (idealized in the point) - that the unresolved state of the dimension problem has something to do with the historical fate of geometrical understanding. In other words, the questions of definition for geometrical forms such as "straight" or "orthogonal" which are left open by Galileo, and their erroneous connection with measurement (or supposed solution through a recourse to measurement), are a second indication that the unresolved state of the dimension problem hangs together with the unanswered question about what the relation is between properties of the world of bodies and the idealized terminology of geometry.

Galileo represented the viewpoint which comes closest to that of the educated layman today. We describe the world of bodies, in some cases with the help of measurements, in a system of three axes which are perpendicular to each other. It is commonly understood that these axes are straight, and that there are no more than three of them. Perhaps a short look at Leibniz might illustrate the degree of certainty which this assumption (by no means self-evident - even for Aristotle) has achieved. In the *Theodicee*, Leibniz uses three-dimensionality, in an entirely different context, as a seemingly clear example for the purposes of elucidation. In a critique of Bayle on the question of how far God could exercise free choice in the creation, Leibniz claims that God, for reasons of geometrical necessity, had no other possibility than precisely three dimensions: for the geometricians had been able to prove that only three straight lines could intersect each other perpendicularly in one point. But where was this proof ever really demonstrated? And what role could this *geometrical* proof play, provided it existed, for the properties of the physical world (which is what this passage from Leibniz's *Theodicee* is all about), unless the definition of the fundamental geometrical concepts are connected, in a specifiable way, to the properties of the world of bodies?

The verifiability of universal historical statements is always a problematic business, since it can hardly be compellingly shown that there has never been a geometrist who has found this proof. We therefore prefer a systematic argument: Is it at all possible that Leibniz's claim is true?

According to the interpretation of modern geometry, anyway, it does not follow from the concepts of orthogonality and straight lines that only three straight lines can stand perpendicular to one another. Which additional premises (perhaps historically older) could come into question, so that any geometrist could have demonstrated the proof that Leibniz claimed? They would have to include some reference to the world of bodies. So one might well assume that this was another one of Leibniz's (not exactly rare) references to accomplishments, of which Kant appropriately wrote (although in a different context): "The famous Leibniz possessed many real insights with which he enriched the sciences, but even greater drafts for such, the execution of which the world waited for in vain."

Kant supposed that perhaps Leibniz had fared as the great "chemists," who "often boasted of tricks as if they were in possession of such, while actually they only persisted in the conviction, trusting in their cleverness, that the performance of these tricks could not fail should they ever wish to undertake them."

SUMMARY

It looks as if Kant's critique of Leibniz would hold true for all previous attempts to provide a basic explanation for three-dimensionality "purely spatially":

1. The general distinction of the triad, from which the triad of spatial dimensions is supposed to then result, originated from the Pythagorean tradition and its belief that the world is ordered by numerical relations. This belief was so shattered by the Greek geometricians' discovery that the ratio of side and diagonal in a pentagon or a square could not be represented by a ratio of integers, that it no longer found followers.

2. To all appearances, by relying on everyday language and mundane practice, ancient geometry annexes those extra-scientific givens according to which, 1) bodies have length, breadth, and height, but not a fourth determinant of that kind. 2) Points are to be seen as the boundaries of lines, lines as the boundaries of surfaces, and surfaces as the boundaries of bodies. There is, however, no fourth type of boundary. At the same time it makes use, without any kind of proof, of the completeness of the enumeration of surface, edge, and corner, with regard to the form of regular polyhedra.

3. The starting point of division or divisibility which gives Aristotle the decisive criterion for three-dimensionality remains unclarified. His continuum theory is developed from one-dimensional constructs, namely from lines and from time; the double divisibility of surfaces and the triple divisibility of bodies are subsequently a logical consequence of the assumption (independent of the continuum theory) that bodies have length, breadth, and height. It is, on the other hand, not the result of an examination of the divisibility of bodies; rather, it is simply a tacit assumption which has always been made, that bodies have three dimensions or extensions.

4. The possible connection between "dividing" and "bounding" is also not sufficiently pursued. There can only be a boundary where there is something to be bounded; thus the bounded object must already be available in order to reasonably speak of boundaries. And in the normal sense of the word "dividing" (referring to the human activity) the bounded object can only be a body. This fact is never made use of. On the contrary, the procedure (even then traditional)

of constructing geometry synthetically - beginning first as planimetry, from the point on, and then, only after this has been completed, to proceed to stereometry - won the upper hand against definition theoretical considerations.

5. In the end, the relation between speech about real, physical bodies and speech about ideal, geometrical objects such as points, lines, and figures remains completely unexplained. Several Greek expressions can be equally applied to real and ideal areas. This may be said in spite of Platonic and Aristotelian conceptions of this relation, because the contradiction of the indivisibility of the geometrical point and the divisibility of the bodies "point," the corner, remains unresolved. Modern mathematics has not picked up on this problem, but has rather programmatically sealed it out of its jurisdiction; it does not claim that a intuitive meaning will be given to those geometrical or topological words to which, for historical reasons, intuitive meanings still adhere, or that they shall, or can, be accompanied by a definition from which such a relationship would become apparent.

Therefore no argument for the triad of dimensions can be won from the previous purely spatial approaches, whether they be numerical speculations, purely geometrical, purely physical, or mixed approaches. Rather, a great number of continual misunderstandings are created, especially in the area of purely spatial statements:

The shabbiness of the definitional foundation of geometry since Euclid's day, whereby some words are assumed to be known, while others are held to be in need of definition (but for which only logically and theoretically incomplete "definitions" are formulated), remains practically unchanged. From the perspective of modern mathematics, an explicit definition of all the fundamental concepts is held to be either impossible, or is seen as a non-mathematical problem.

The thesis that there are only three straight lines, perpendicular to one another, which are suitable as coordinate axes for describing the location of bodies, has only been claimed, never proven - neither as a geometrical statement, nor as a statement about physical bodies such as straight rods.

The situation has always been such that one merely accepted the "necessity" of three-dimensionality, in the sense of everyday understanding, since one did not know of an intuitive alternative to the triad of orthogonally paired rods or planes. (This ignorance was supposedly supported by the claim that more knowledge or proof is quite simply impossible.) Or else it was alleged that others have given, or could give, the proof or something similar.

One can see a certain acknowledgement of this theoretical destitution only in that it is often attempted to show the three-dimensionality of space with stronger, additional assumptions, which only makes sense as an undertaking if the weaker, purely spatial ones are not yet sufficient. Therefore we now turn to the attempts to demonstrate three-dimensionality with the help of aspects of motion and direction.

CHAPTER TWO

Grounding Three-Dimensionality in Motion

PRELIMINARY REMARKS

It is not known who first came up with the idea of trying to deduce three-dimensionality from a discussion of the motions of bodies or of geometrical figures. It is rather improbable that the beginnings of such an idea go back to ancient geometry where on one hand, as mentioned above, the three-dimensionality of spatial forms was taken for granted but, on the other hand, it was accepted by way of proof in plane geometry to flip a figure out of the plane in order to bring it to cover a mirror-congruent figure. In the first place, there was (as witnessed by the above-mentioned definitions of Euclid) obviously no need, apart from the spatial distinction of length, breadth, and height, for an explanation of three-dimensionality. Second, according to the ancient view, the concept of motion belonged in physics. Mathematics was, in contrast, the science of temporal immutables.

One could, therefore, view the text "Concerning the Ultimate Foundation of the Differentiation of Regions in Space" by Kant (1768) as one of the first texts to consider this problem, although clearly only with a modern interpretation which ignores Kant's original intent. For Kant this tract, directed against Leibniz, was namely "to find an evident proof, ... that absolute space has a reality of its own independent of the existence of all material." So it was intended to prove the existence of an absolute space, in Newton's sense, against a relational interpretation of space, and Kant believed to have found his own argument in favor of this, namely the existence of "incongruent counterparts." (Incongruent counterparts are, for example, a left hand and a right. They cannot be superimposed by shifting them; intuitively speaking, the left glove does not fit the right hand and vice versa.)

However, Kant does not use this consideration as a demonstration of three-dimensionality. Rather, he expressly assumes the triad: "In corporeal space, *because of its three extensions*, three surfaces can be imagined which intersect each other orthogonally." The modern argument, according to which only one- and two-dimensional forms can be converted to their mirrored counterparts by motion through higher-dimensional spaces, was the first to deduce the absence of a fourth dimension from the impossibility of doing the same thing with three-dimensional forms in the sense of the Kantian "incongruent counterparts." Therefore, we will use the Kantian text as an occasion counterparts." Therefore, we will use the Kantian text as an occasion to discuss this modern argument.

While this argument deals with the *motion possibility of real bodies*, another group of authors attempts to extract an argument for three-dimensionality from the discussion of *imagined motions*. William K. Clifford attempts to determine surfaces, lines and points as boundaries for imagined motions through real bodies. Paul Natorp observes the rotation of line segments and planes, and believes to be able to demonstrate three-dimensionality from the impossibility of rotating space as a whole. Adolf Trendelenburg advances a variation of the widespread view that, through motion, points become lines, lines become surfaces, and these, in turn, create spaces when moved. For him, points "stretch" themselves into lines, lines stretch into surfaces, etc.; it is then stated that a body cannot stretch itself into anything other than, again, a three-dimensional body. We will deal with these approaches under the title "imagined motions."

First of all, it must be remarked that the foundation of three-dimensionality based upon motion *cannot be a geometrical problem*. Although the concept of motion is a useful metaphor for geometry, which one certainly need not renounce, it is also only a metaphor for a type of "mapping" or for a group of constructions.

A plane triangle, for instance, is "moved" (by means of translation) in the following way: three equally long line segments are constructed, beginning at the corners and going in the same direction (that is, parallel). The end points of these segments then form a new triangle, congruent to the first, which one can interpret as being the original, which "moved" itself along the path of the three parallel segments. Rotation motions with circular arcs, as well as motions through "spaces of higher dimensions" (for example, flipping a triangle over to convert it into its mirror-congruent counterpart), are analogous to this.

Such "motions" of geometrical figures provide no argument for the thesis that spaces whose dimension number is one higher than that of the moved figure are required for the conversion of a figure into its mirror-congruent counterpart. For this "motion" is now only a metaphor and is replaceable by another metaphorical way of speaking, namely that of "mirroring." Since we are going to pick up on this issue once again in the discussion of Kant's treatise, the following reference may suffice here: When a (non-symmetrical) triangle is, by means of reflection, converted into its "incongruent counterpart" this is done through the following construction: one arbitrarily chooses any given straight line lying outside of the triangle on the same plane, then one draws perpendiculars to this line from the corners of the triangle and, finally, one doubles the length of these perpendicular lines. The end point of these perpendiculars are the corner points of a new triangle, mirror-congruent to the first. (See Fig. 1) This construction does not leave the plane. Analogous is the reflection of a three-dimensional, oriented form in a mirror plane. As in the case of a screw threaded to the right, all points are converted through reflection (again with the construction of doubling the perpendiculars at the mirror plane) to the corresponding points of a left-threaded screw without needing an extra dimension. The (supposed) argument in favor of three-dimensionality based on the impossibility of converting an oriented body (like the right hand) by motion into its mirror-congruent counterpart (the left hand), must then refer to something other than some kind of geometrical state of affairs. This is true even when truly practicable operations with compasses and rulers (i.e., with bodies) are what is meant by "construction," and not simply intuitively "imagined" operations which are only carried out with calculations using spatial coordinates.

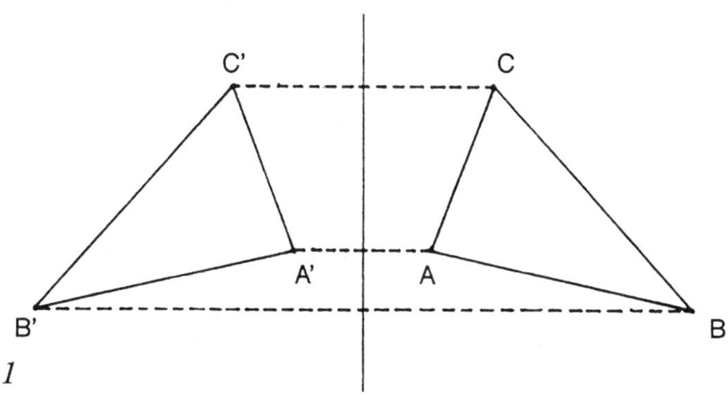

Figure 1

KANT AND THE INCONGRUENT COUNTERPARTS

In the above-mentioned text of 1768, Kant attempts to demonstrate the existence of absolute space by maintaining that it is impossible to convert an (oriented) three-dimensional form, like a human hand, into its corresponding counterpart by motion (as defined by reflection). But because all relative determinants of the form of a hand (like lengths, angles and the position of the parts relative to each other) are the same for the right and left hands, in Kant's argument, an extra, absolute factor must enter into it which accounts for the difference between the right and left hands, and that is the absolute space.

It is indeed peculiar (and can truly only be ascribed to the abstinence on the part of many philosophy historians from any kind of systematic thought) that this short text of Kant's - only seven pages long in the Akademie edition - has continually been granted great attention, and is discussed to this day. Because for an impartial observer, not already paralyzed by the historical authority of Kant, this text proves to be poorly thought out, full of mistakes and questionable assertions, sloppily formulated and far from convincing.

1. To explain the difference between the right and left hands or between right and left-handed screw-threads that are otherwise the same, Kant begins with the assertion: "When two figures drawn on a plane are equal and similar to each other, they cover each other. *Only with physical extension, or else with lines and surfaces that do not lie in one plane*, it is often completely different." (My italics. P.J.) This is false, as any geometry layman can also instantly see. Any given triangle, as long as it is neither equilateral nor isosceles, has a "counterpart" which has equal sides and angles, but which cannot be superimposed over the first by a shift in the plane. For a demonstration of this, one needs only take a commercial drawing triangle with 30, 60 and 90 degree angles, lay it on a drawing surface and trace around it, first on one side and then on the other side. And yet, the Kantian criterion that they be the same regarding all determinants of "relation and position (of their) parts to each other" also applies to this triangle.

2. The definition of "incongruent counterparts" (of which the two hands are examples) as "the image of an object in the mirror" can also be applied to two-dimensional, plane figures; yes, even for one-dimensional, straight figures, so that Kant's argument for "the possibility of wholly similar and equal, and yet incongruent spaces" also is true for those of only one or two dimensions. But it would be extremely vexing for Kant's view to have to accept the existence of absolute planes or lines as well as the existence of absolute space because the Newtonian laws of motion are invariable concerning rectilinear translations. Thus, the rectilinear translation motions are not to be regarded as motions in a *one-dimensional*, "absolute" space. Besides, the "bucket experiment," with which Newton wanted to demonstrate the existence of absolute space, is impossible in a one-dimensional space. Hence, Kant's argumentation has already failed in its main objective.

3. Kant says: "The outer surface that bounds the physical space of the [right hand], twist it and turn it how one please, cannot act as the boundary of the other [left hand], hence, this difference must be such as is based on an inner foundation." Here Kant makes an important corollary, which he does not mention, and whose significance he does not truly understand. Every layman knows that when one "turns " a right-hand glove, that is, turns it inside out, it will indeed fit the left hand. "Twisting and turning," then, must not include inversion. The

admissible motions must be "*rigid*" motions instead, which is to say that the moved body may not be bent, stretched, or compressed. The fact that in the act of turning a glove inside out, the glove must be stretched and compressed can be seen from the following example. One presses a portion of a rubber ball inward. In doing so, a more or less circular area is changed from a convex to a concave form. Along the way, the rubber surface must pass through the form of, approximately, a plane, although the intersecting plane through a sphere is obviously a smaller surface than the spherical cap above it. Thus, the rubber ball must be compressed in this area. This is also the case for every finger of a glove.

Kant does not, however, mention the condition of rigidity of motion, nor does he have a criterion for it here or in any other text. Furthermore, none of his contemporaries or predecessors knew of such a criterion. One must first know how to practically check the invariability of a form during motion before one can arrive at such a criterion.

4. The empirical evidence that Kant introduces as "*Kennzeichen der Naturzeugung*" (characteristics of natural generation), which is supposed to provide him with an argument against a simple definitional-relational differentiation of spatial orientations, is highly dubious. How could Kant have possibly known that "the hair on the crown of every person [is] ... whirled from the left to right." Or that the direction of the spiral of a snail shell is independent of the direction of the path of the sun, and is thus the same in the northern and southern hemispheres? And it is the sign of a particular program of upbringing (as far as it is at all true) that one writes with the right hand "everywhere," and not a "characteristic of natural generation."

5. Kant distinguishes "the regions in space" by reference to three planes, each orthogonal to the others, which are situated in the human body. We orient these planes in relation to ourselves with the words "above/below," "left/right," and "front/back." Kant points out, and rightly so, that we can only then distinguish the "world regions" and draw accurate star charts, when we take this sense of orientation into account. For example, when retaining the North-South direction, we do not make the mistake of switching it for the East-West direction. We can do this relationally, namely by reference to our bodies, and the learned difference between right and left. Kant also correctly sees that we can give clear descriptions of some spatial forms (precisely the insufficiently symmetric or the mirror-invariant ones) *only in relation to orientations or directions*. But he does not admit direction itself to the determinants (like those of hand). Thereby, the reference to our bodies for the normal linguistic means with which we specify orientations is precisely the solution to the definitional problem that Kant wanted to solve by reference to an

absolute space. In fact, we cannot "absolutely" define any measurement unit (like that of length); instead, we must refer to a standard, a unique thing (like the Paris standard meter, the earth's meridian, or a particular yard or a foot). This obviously provides no more evidence for the existence of absolute space, than the reference to standards does for an unequivocal definition of "left-handed" or "right-handed."

Therefore, we will now leave Kant's text of 1768 and, after having already shown that this is not a *geometrical* problem, will examine whether or not the *mobility of bodies* produces an argument for three-dimensionality.

THE ANALOGY ARGUMENT AGAINST THE FOURTH DIMENSION.

A current argument for the three-dimensionality of empirical space (hereafter referred to in short as the "analogy argument") is approximately the following: let us assume we that have a one-dimensional, a two-dimensional and a three-dimensional form, like an arrow, the letter S and a helix. Now the one- and two-dimensional forms are each converted to their own reflection by a half rotation; the arrow is rotated 180 degrees in the plane and the letter S is also flipped 180 degrees through space. These half rotates each require a "space" whose dimension number is one higher than that of the rotated form itself. *By analogy* (hence, the "analogy argument"), the helix would have to be flipped through a four-dimensional space in order to be converted to its mirror image. But, intuitively, we cannot imagine such a rotation motion, nor do we have any experience of the kind in which a helix (like a metal screw) is converted to its mirror image by rotation; from a right-handed to a left-handed screw, for instance. So much for the usual, uncritical formulation of the argument.

It already becomes clear at the first glance that this argument can only be understood in its intention when a few extra conditions, which are not explicitly mentioned, are accepted.

First of all, it must be known (i.e., defined) what the reflection of a respective form is. In ordinary, extra-scientific language usage, the most familiar "mirror writing" is obviously that which one obtains by looking at normal writing in a mirror. "Mirror image" is thus taken from everyday practice in the use of plane mirrors and optical reflections (which is, incidentally, also the case with Kant).

A second silent assumption of the analogy argument is that the problem is not to be understood as purely geometric or as a construction example. This assumption is even made in two ways, namely, naively and in a terminologically reflective way.

In a naive sense, it cannot be a purely geometrical problem because the concept of mirroring is not restricted to the geometrical construction of the mirror image, but makes the additional silent assumption that physical objects like a drawn arrow, a written S and a real helix, made of wire, for instance, are seen in a mirror. If this additional assumption were not fulfilled, an additional condition concerning the position of the mirrored form to a mirror would have to be specifically mentioned in order for the analogy argument to apply to the two-dimensional case, for instance. We always see the real, written letter S as a mirror image in a mirror, regardless of the angle the mirror and the letter have to one another - as far as we can see it at all. However, the geometrical construction of the mirror image does not lead to a letter S in mirror writing in the case of a parallel position between the mirror and the letter (see Fig. 2).

In a terminologically clarified way, again, it becomes clear that the analogy argument applies only to real forms, hence bodies, and not to abstract lines. (As it is, it is not even to be considered here that the letter S and the helix as abstract lines are, of course, only examples of one-dimensional forms; hence the imputed condition of two- and three-dimensionality is only then fulfilled with the additional assumption that they are understood as, respectively, a plane form and a figure on a cylinder surface.) Purely geometrically, the analogy argument is simply false. This can be seen in the following: the construction of the mirror image, with the proper position of the mirrored figure relative to the mirror plane, namely the drawing of a perpendicular to the mirror plane from every point of the figure, and the doubling of these perpendiculars, is already a point-by-point rotation, so that it can hardly be surprising when half rotating a figure converts it into its mirror image. After all, doubling the perpendicular as a geometrical task of construction is solved by sticking dividers in the foot of the perpendicular at the mirror plane and then conveying the distance between figure and mirror to the other side of the mirror plane. In other words, the generation of a mirror image as a construction task is nothing more than a point-by-point rotation of a figure 180 degrees; although this is a meaning of the word "rotation" which differs from everyday language usage. In everyday speech, what we mean by "rotation" is the motion of bodies around a rotation axis, which is naturally a straight line, hence a one-dimensional form. For the geometrical construction of a mirror image, however,(under the above-mentioned additional conditions of position) the "axes" are, in the case of an oriented line segment, a point; in the case of a S-line, a straight line; and in the case of helix, a plane. In precisely this way, the (point-by-point) "rotating" of a helix

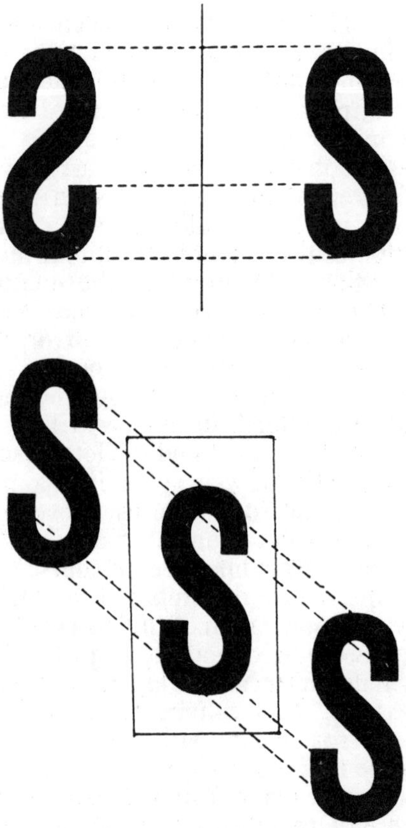

Figure 2

around the mirror plane becomes an inversion which reverses the direction of the helix's spiral. Clearly, for two and three-dimensional instances, the construction of mirror images in no way requires a "space" for the construction which has a higher dimension than the figure to be mirrored. Hence, the intention of the analogy argument can only then be understood if it refers to real forms, i.e., bodies. But can the analogy argument be validly formulated as an explicit empirical argument without logical contradictions?

To test this the following bodies will now be considered as examples of one-, two-, and three-dimensional "forms": a bullet with one flat and one pointed end as the one-dimensional example; a lead template such as cobblers use for cutting leather as the two-dimensional example; and a metal screw as the three-dimensional example. For these three examples, the following facts are accepted as undisputable:

In order to reverse the oriented bullet (oriented, as used here, means that the ends are distinguishable from one another, e.g., the point and the flat end of a bullet) in its position relative to a gun barrel, itself oriented, as happens in the case of the mirroring of the drawn arrow, the bullet must be taken out of the barrel and rotated 180 degrees on a cross axis. In order to reverse the template for the cutting of leather for right-hand gloves to its mirror image (for the left-hand glove) one must lift it out of the plane of the cutting table. Meanwhile the leather, itself oriented with regard to outer and inner sides, must, of course, remain lying on the table. There is no rotation for the screw, however, which would similarly result in a mirror image screw. That is to say, given a right-handed screw and a left-handed nut of the same gauge, no rotation of the screw relative to the nut could bring the two to fit together. This "empirical finding" forms the basis of the analogy argument.

It is nevertheless rather easy to see that, regardless of the empirical facts, no argument for three-dimensionality can be won from this. It would have to be shown that in the one- and two-dimensional instances of rotating, something *succeeds*, which we know from experience *fails* for the rotating of a three-dimensional form. But this is not the case. Obviously, the bullet, template and screw all behave in exactly the same way. All three are bodies, on which we can mark three directions which do not lie on one plane, and in all three cases, when rotating on an axis they behave in exactly the same way. The illusion that the analogy argument creates rests upon the fact that, in the cases of the bullet and the template, we focus our attention on only one or two directions on the respective bodies, and ignore the other extensions. The bullet and the template are no more converted to their "mirror images" or "incongruent counterparts" (Kant) by a rotation than is the screw.

A last attempt to rescue the analogy argument could be formulated in the following way: The generation of "mirror images" through the geometrical construction of a mirror plane makes use of the point-by-point rotating of a figure. This is possible for geometrical figures, that is, abstract forms, but not for real bodies, which should not be broken up into their smallest possible components for the purpose of a rotation. Hence, it is something *specifically corporeal*, namely the inner composition of a body's parts, which prevents point-by-point rotation and the consequent conversion into a body of the mirror image shape. The above-mentioned inversion of a right-handed glove into a left-handed one as the realization of a point-by-point rotation is also not a counter-argument here, because the glove is, here again, merely seen as a two-dimensional form which only exchanges the outer and inner sides when inverted. If, however, one considers the copy of a right hand, for instance, made from an elastic material, no inversion to the left counterpart could occur without the material, be it ever so elastic, ripping apart in the process. Thus, it is truly something about the three-dimensionality of bodies, which is only accessible to us through experience, which makes it impossible for us to convert bodies through motion into their mirror image shapes.

But this rescue attempt is also doomed to failure because the problem already disappears when the linguistic means which are indispensable for a gapless, contradiction-free formulation of the problem are elucidated, quite independent of experience. This can be shown without great mathematical or logical effort by looking at the previous argumentation.

We can, with the help of two casting molds (one for each hand), convert a fluid, (water, for example) from a body in the shape of the right hand to a body in the shape of the left. This is surely as undisputable as the above-mentioned, experiential facts for the motion of the bullet, the template and the screw. Thus, it cannot be attributed to the corporeal aspects or three-dimensionality of bodies, which are naturally also present in fluids, when the conversion to the mirror image form does not succeed. It must be attributed instead to the criterion of form constancy. Obviously, the decisive dividing line here is between solid and fluid, to which it must be said that "solid" need not be limited to meaning "rigid" (as in the case of the metal screw), any more than "fluid" must mean a thin liquid like water. The decisive dividing line is rather to be found between more or less elastic materials on the one side, and more or less plastic materials on the other. Accordingly, the cast of a hand formed of a highly elastic rubber would still be considered solid, and the cast of a hand made of molding clay would count as an example of a plastic body. (In the example of the rubber hand, which, of course, can be bent so as to have apparently changed its orientation, the preservation of the sides - the back of the hand and the palm - comes into play; with a suitable bending of the hand, the fingernails of the one hand would come to lie over the fingertips of the reverse oriented hand.) But what does this difference consist of?

What is meant here is obviously a form constancy which no longer allows for a gradual transition from solid to fluid or from elastic to plastic. Terminologically clear, and of high importance for geometry, technology, and physics, is the definition of form constancy and the rigidity of bodies, such as is expected of measuring instruments for spatial magnitudes. *This* determination is, however, defined as a special immutability in transport, translation motions, and rotation motions. In the simplest cases, a body is considered rigid, when in translation of all of its parts along parallel line segments of equal length, it is converted to itself. In rotation, bundles of circular arcs appear around the same line, i.e., a one-dimensional axis, instead of the bundles of parallels which describe the paths of the body parts in motion. In contrast, the "parallelism" of the bundles of parallels or circular arcs is not maintained for plastic materials, in which the parts can freely change their relative positions to each other. In logical terms, the concept pairs of solid/fluid and elastic/plastic are contrary in regard to the properties which are interesting here: what is solid is not fluid, and vice versa; and what is elastic is not plastic, and vice versa.

However, with this elucidation of the concepts the analogy argument takes on the following form: Can a solid (or elastic) body be moved such that it undergoes a shape-change as defined for fluid (or plastic) bodies? Or, to put it more simply, can a body be solid (elastic) and not solid (not elastic) at the same time? This is clearly impossible for logical reasons, because the corresponding characterizations are logically inconsistent. When the description of a situation already shatters on an inner, logical contradiction, one need search no further for a practical realization of that situation, or to strain the experiential foundations - any more than it would make sense to try to resolve through experiment with ruler and drawing compass whether it is not, perhaps, indeed possible to construct a triangular circle.

Thus, the analogy argument clearly has absolutely nothing more to do with the dimension problem. We still speak of *bodies*, to which we attribute three dimensions, in the determination of the concept pairs of solid/fluid and elastic/plastic, but this is not decisive. Analogous differences can also be carried over to abstract forms with other dimension numbers.

THREE-DIMENSIONALITY AND ORIENTATION

When a riddle which is formulated in simple words ("Why can't the left hand be converted to the right by mere motion?") is made to disappear by a critical analysis of its linguistic means, this often leaves laymen, as well as specialists, feeling unsatisfied. The suspicion remains that there must be, apart from all terminological criticisms and attempts at precision (against which one may, perhaps, have no objections), a core of the problem which lies outside of the realm of language and definitional hair-splitting, and which is unaffected by linguistic criticism or linguistic improvements.

Such a skepticism against linguistic criticism is suggested by the analysis of the preceding section, according to which the analogy argument either cannot be formulated purely geometrically (because all definitions are valid for any dimension number) or is quite simply invalid (because the figures of any dimension number can be converted to their mirror image by a point-by-point turn). This is also true for the refutation of the analogy argument for real bodies. In the end, all the criticisms against the analogy arguments - purely geometrical as well as real instances - follow from additional definitions. Perhaps these are the cause for the premature disappearance of the analogy argument as a problem.

Finally, we know from everyday life, independent of all definitions, that the left shoe does not fit on the right foot, that the right-handed screw does not fit in a left-handed nut (which would otherwise fit the gauge), not even when one attempts to screw it in from the other side of the nut. In short, a right-handed screw does not become a left-handed one by turning it around, etc. Might it not, in fact, be a problem of "being" and not just a terminological "language problem" which is hidden behind the problem which Kant deficiently formulated? Such suspicions could receive additional support, in this case, by the fact that the problem is not grounded within the realm of geometry as an exact, definitionally sound science, but rather upon a possibly problematic application of geometrical language to real bodies. Or, briefly: cannot the old problem, at the heart of which lies the unshakeable certainty of our everyday experience, be reformulated so that it has stability in the face of linguistic criticism?

We will examine these suspicions once again here in order to show that they have, in fact, a cause which is to be found in the matter itself. This cause, briefly stated, is to be found in the fact that an unambiguous determination of the words "left" and "right" (in the sense of left-handed and right-handed) is only possible through reference to real bodies. Thus, one needs a distinguished standard body to which one can refer when one must, for example, identify a screw as right-handed or left-handed, while this is not necessary in order to determine whether two differently oriented screws act as mirror images to each other. This can also be expressed as follows: being "oriented," that is, non-symmetrical, and therefore having a mirror image counterpart, is a purely geometrical property of a form, while exhibiting a *certain* orientation is the property of a form which represents a real, or physical, quality.

If one looks once again at the discussion of the analogy argument from the previous section, what is striking is that only so-called "form-theoretical" words appear in the elucidation of the geometrical metaphors "motion" and "mirroring." That is, such words as "plane," "straight," "perpendicular," "circle," "half-turn," and "doubling the perpendicular" are used. All of these expressions are invariable with respect to magnitude; that is, for the way in which they are applied, they are exhaustively determined without reference to a standard of magnitude or measurement of length. In the philosophical discussion about the foundation of geometry one calls such geometrical concepts "prototype-free" definable or reproducible concepts. What the definitions of fundamental concepts of geometry like "plane," "orthogonal" and "straight" have to do with the practical manufacture of spatial forms on real bodies will be explained in Part II of this book. It is sufficient here to realize that one moves within the realm of geometry, which manages with concepts which are determined by pure form theory, i.e., those for which no reference to a distinguished standard body (like the Paris standard meter as a unit of length measurement) occurs.

The opposition between symmetrical and non-symmetrical can be defined with pure geometry. Symmetrical figures, like the unoriented straight line segment, the circle, the sphere and, of course - in reference to particular planes - the cube, the cylinder, etc., are such that their mirror image cannot be differentiated from the original. (It is geometrically irrelevant that we are normally accustomed to laying the mirror plane inside a figure, through some "midpoint.") Non-symmetrical, then, are those figures for which translation and rotation (synonymous with mirroring) do not lead to the same result. The "existence" of pairs of mirror image figures is guaranteed by geometrical definitions and has nothing to do with the world of bodies.

If one assumes this knowledge to be at one's disposal, a task can immediately be formulated for whose solution this knowledge does not suffice. One imagines a box full of screws, which are equal in all possible ascertainable properties, with the lone restriction being that both orientations of the screw thread are to be found. And now the task of taking a right-handed screw out of the box is imposed. In order to insure that this task is as free from the use of a standard object as possible, we will imagine that a Person *A* requests of a Person *B*, by telephone, that he pick a right-handed screw out of the box, but that Person *B* does not know the words right-handed and left-handed; on the other hand, all geometrical words are known to Person *B*. Person *A* then has the task of explaining over the telephone how right and left-handed screws are to be recognized. (Person *B* is already aware of the fact that they have a mirror image relation to each other.)

The solution to this task is to be found in an appeal, over the phone, to an appropriate standard of comparison. In English, one normally makes reference to the "clockwise" direction for this purpose. This, in turn, is oriented on the direction of the sun's apparent motion across the northern hemisphere. These, then, are the factors which are exploited in the solution: a person living in the northern hemisphere always has an orientation, and if the call should be placed to Australia, one can refer to a commercial clock (which the cultures of the northern hemisphere forced upon the inhabitants of the southern hemisphere a long time ago). However, the caller, trusting that his telephone partner has the normal arrangement of organs, could also require him to refer to the side of the body his heart is on, and to draw a conclusion about the left and right hands from this.

This example should graphically portray that the *identification* of a screw as right-handed only succeeds through a reference to a *body*. Hence, while the definition of the relation between right and left succeeds geometrically, the unambiguous determination of these two words is only possible using prototypes.

The uneasiness which was spoken of - that the analogy argument can be made to disappear through linguistic criticism - can now be clearly understood: our intuitively reliable experiences of the fact that we can no more switch left and right than we can switch left and right shoes has nothing to do with the mirror imagery of such pairs of objects, but rather becomes a factor only when an unambiguous identification of one of these two orientations is demanded. It is confirmed that the analogy argument as proof for the three-dimensionality of space is false. In contrast, when we want to make a clear choice in the realm of bodies, which relate to each other as mirror images in reference to their form, we always need a reference to a particular, distinguished standard example for a spatial orientation. And since the human body already exhibits certain symmetries, like those of the hands and the feet, which nonetheless must be clearly differentiated in daily dealings with, say, pieces of clothing, our intuition does not deceive us when we attribute this accomplishment, not to geometry, but to the world of bodies.

The fact that the identification of right and left does not succeed in one or two dimensions is easily demonstrated. For instance, when the letter *S* is drawn on fully transparent plastic, the "correct" letter and the mirror writing letter can only be differentiated when it is said "from which side" the transparency is to be viewed. But the writing plane and the prescribed viewing direction present us, once again, with a three-dimensional reference system. The one-dimensional instance is analogous. In geometry (in which the expression "viewing direction" obviously does not occur) this state of affairs was conceived as the difference between shifting the letter *S* in the writing plane and lifting the letter out of the plane.

IMAGINED MOTION

Three authors (Paul Natorp, Adolf Trendelenburg, and William K. Clifford) were previously mentioned, for whom motion plays an important role in providing a basis for the three-dimensionality of space. What this means, on the one hand, is that, for these authors, it is motion which recommends - or even compels - the acceptance of the triad, and, on the other hand, that these three authors are exemplary representatives of particular types of arguments. A complete treatment of all historical approaches which attempt to prove three-dimensionality on the grounds of motion is not the goal here, however.

A type of consideration which seems especially obvious can remain unexamined here: one could ask how many dimensions space must have, minimally and maximally, so that a *point-body* could *move* from any place in space to any other place. Because the answer clearly depends on how one wants to talk about the potential position of bodies in space, and with what right one idealizes bodies as point-bodies; why one believes it to be possible, for example, to be able to ignore the form and magnitude of real bodies and, thereby, their spatial orientation. Finally, it is only a purely spatial question, in which it is a matter of determining the position of a point "in space," i.e., relative to another point, one or more straight lines, or planes. If the position is unambiguously describable, then the beginning and end of a motion of a body which has been idealized as a point are also determined. An approach is only included among those to be discussed when something specifically new arises from motion, like the continuity of line segments, or of space as a definitional consequence of the continuity of motion. And, indeed, one of the above-mentioned authors, William K. Clifford, can be called a representative of this approach to the problem.

Likewise, it is obvious that "imagined" motions must have some reference to the world of bodies. We have already discussed that the "purely geometrical" motions are mere metaphors for geometrical constructions.

As has already been the case in other passages, I must confess that I do not know who was the first to come upon the idea of idealizing real body motions as abstract point motions. When the curve of pursuit, or "dog curve," was defined in antiquity (the Latin label is *tractrix*, or equitangential curve, i.e., those curves whose tangents always have a constant length with a straight line from the point of tangency to the intersection point) a very intuitive physical situation was the basis for this definition of a curve as the path of a point (dependent on the motion of another point). A person holds a dog (which resists motion like a stone) on a taut leash, and now begins to move at exactly a right angle to the taut leash, dragging the dog behind him. The dog will then describe a "dog curve" on the ground, that is, his course will approach the straight path of his master asymptotically. The geometrical treatment of this problem is an idealization of a procedure with moved bodies and must primarily be directed by the image that the point which idealizes the person "moves" along a straight line. Then of course the tractrix can be defined as the geometrical position of all points which form the point of tangency of a tangent of a constant line segment with the intersection point of a prescribed straight line. The image of motion no longer appears here, but the name of the curve nevertheless indicates the kinetic original context of the geometrical problem of definition.

The most abstract attempt to deduce the three-dimensionality of space from the concept of motion doubtlessly originated from the Marburg philosopher, Paul Natorp.[8] For the neo-Kantian Natorp there are "no objects given to thought, before they are created by thought," which is why "until now, nobody has been able to specify the variety of objects from which the concept of dimension is supposed to result independent of the presupposition of dimensions. Naturally, it is the dimensions of space that one has in mind here." In a paragraph titled "The Three-Dimensional Euclidean Space," Natorp wants to "determine the exact law and the possible upper boundary for the ... expansion of the concept of space," after "the majority of directions and dimensions," that is, the circumstance that more than one direction must, "exist perforce mental necessity itself." (We will not go into Natorp's line of thought according to which the existence of only one direction or dimension is held to be impossible; we will only consider his argument for the triad as upper boundary here.)

Natorp alleges "a single base direction ... through any two elements (O and 1)," to which all other directions in space should refer. However, no upper boundary for the number of directions or dimensions results from this. This is not just a matter of mere "*Denkbarkeiten*" (conceivabilities), but is supposed to deal with "the possibility of ontological determination." In other words Natorp is not concerned with purely mathematical determinants, rather with those whose applicability to the world of bodies must be able to be shown or given.

With a digression on Kant, Natorp gives the reasons why there must be an upper boundary to the number of directions at all, to the extent that one does not stay with "pure" mathematics, and instead wants to declare "determinants" for existing things - "directions" or "dimensions" are such determinants. But if there were an infinite

8 P. Natorp (1910) pp. 263-309

number of these, then one could not determine anything real with them. Hence, the dimension number must be limited in such a way "as is necessary and sufficient to produce a single, hence closed, connection of spatial determination, homogeneous and continuous." According to Natorp, then, it is the continuity and homogeneity of space which demand the dimension number three. And then comes the assertion, short and sweet: "The proof is simple."

It amounts to this: that an oriented line segment (0,1) can be continuously turned only to a maximal directional difference of the opposite direction (1,0), and any further turn of the segment would lead back to the original. This continuous turning generates a plane. Proceeding from the base direction (0,1) this plane can, itself, now rotate around the base direction and, in this manner, generate the three-dimensional Euclidean space.

> But our principle of deduction does not lead to more dimensions. According to our principle, new dimensions are to be introduced if, and only if, they are necessary in order to convey given directional oppositions in continuous transition.

In contrast, the fact that a plane can rotate around a fixed axis in two different directions does not require an extra dimension, because this rotation continuously passes over into a opposite direction when one now additionally gives the rotation axis of the plane a half turn to one of its perpendicular axes. Since the space of Euclidean geometry is homogeneous, that is, since every last one of its points is undifferentiated with regard to every construction, the intersection point of both rotation axes just mentioned can also lie in any point in space, and every directional difference can be determined from any point as well.

It is helpful for the understanding and judgment of the Natorpian approach to consider the end of this paragraph, where Natorp briefly mentions (n-1)-dimensional, non-Euclidean spaces (like the sphere on which there are no parallels) and stresses that non-Euclidean geometries which are free from logical contradictions are "conceivable, ... exactly representable conceptually, and mathematically calculable in all of their basic properties and consequences," but that as merely mathematical spaces - as opposed to Euclidean space - they would not come into question as "conditions of existence."

One of the two premises on which Natorp bases his argument becomes clear at this point: it is only *Euclidean space* which can be acknowledged as an existence condition. Euclidean geometry is the only theory which formulates this condition of existence. In reference to the Natorpian text, however, this means that properties of bodies must be used for support in some way: it does not follow from a "purely mathematical concept of plane" (i.e., one which is not already defined as a cut through a body or as a plane body surface) that with a rotation around a straight line which lies in the plane and a rotation around a straight line which lies perpendicular to the first that all directions "of space" can be reached (as in a "space" with a dimension number larger than three). Only a *body* can, with a direction which is drawn upon it (through two marks O and 1), be brought to any direction prescribed on another body through two such turns. It remains an open point with Natorp, however, as to how the connection between the "description" of such spatial properties of bodies with the definitions or theorems of Euclid is to be viewed. In other words, how is one supposed to recognize which geometrical statements are responsible when a theory formulates "conditions of existence."

So the knowledge that every spatial direction can be achieved by proceeding from any other given direction through two (simultaneous, or in sequence) executed turns around a pair of orthogonal axes, is only valid in a three-dimensional space. Thus it is a *petitio principii*, or it must first be won as knowledge about corporeal properties through a logical-definitional union of ideal, geometrical language and real, descriptive language for bodies.

Without such proposals (missing with Natorp) the argumentation hangs in the air, in a literal sense. When a base direction (0,1) is supposed to be taken as a reference, then the question is what it means, more precisely, "to take" such a direction. How does one determine such a direction without reference to other bodies or references? Even with every effort to follow the Kantian tradition of pure thinking, nobody can "think" an oriented line segment with a direction, without any reference to something else; to some coordinate axes, a nest space, or an imagined space (oriented on visual perceptions and reference to our body axes). If, however, one tries to fill this deficit, one does either a *pure* geometry, where, in accord with Natorp, there are no reasons for the restriction to three dimensions, or one does *intuitive* or applied geometry, and so has to finally speak of the properties of bodies and about artificially introduced or generated characteristics, such as the distinction of two places on the body through the use of marks. This latter possibility, however, would have to be described with an appropriate vocabulary, or to first be made describable, in order to not suddenly turn into an empirical proposition about the possibility of turning a body. In the framework of the Natorpian approach, however, this problem must not, under any circumstances, become an empirical question, since the necessary objects must first be "produced ... through thinking."

To the extent that one extracts the Natorpian proposal from the fog of neo-Kantian transcendental reasoning, one can at least see a remarkable proposal therein: one proceeds from any given body, say a stone, and marks two spots on it; now one holds these fixed relative to a second body. One can now speak of rotation around these two spots relative to a second body. By repeating the procedure it is now possible to rotate this second body, relative to a third, in such a way that the two rotation axes, now defined, form a right angle together. One remembers, for instance, Bishop Cartan's invention and the "Cardanic suspension" of a ship's compass. It is then a well-defined question whether by superimposing two rotations with axes which are at right angles to each other it is possible to replace all other rotations. If one adds a motion along the radial direction, so that two such rotations, plus translation away from the rotation point, equals three motions of which none can be replaced by one of the others, then there is, in fact, a motion argument for three-dimensionality here. It would surely be fair to call this a "Cardanic-Natorpian" solution.

This solution would, however, be lacking the extension to such questions as: To what degree is a cuboid (as a static form, realized in, for example, a crate or a brick), with length, breadth and height, three-dimensional? What does the ability of bodies to rotate have to do with their spatial form and extension independent of all (rotation) motion? And the criticism remains that to build a Cardanic suspension, we must be able to lay the axes "correctly," that is, to realize the plane and the right angle. Hence, we are forced by practical considerations to fall back upon a technical realization of purely spatial three-dimensionality.

The criticism against the Natorp's considerations can be summarized as follows: talk about the superimposition of rotations already makes use of geometrical concepts which, for their part, when applied to the world of bodies, already comprise a decision about the dimension number.

Adolf Trendelenburg, whose critique of Hegel was quoted above, focuses on a totally different type of motion, which is not confined to being a mere motion-oriented metaphor of a purely spatial state. He writes in his "Logical Investigations" (1842):[9]

> Why does space have three dimensions and not more and not less? ... If we imagine ourselves at the generation of a mathematical body and we allow the three dimensions to come into being in this body, then we run into the actual root of the difficulty. The point which we tentatively set strives beyond itself and extends itself in this manner to a line; the line moves out of itself and widens itself in this manner to a surface; the surface describes through its motion a body. It is a seeming contradiction that the point goes out from itself;

9 A. Trendelenburg (1862), pp. 224-226

> this it is, but nothing other than motion itself, in the beginning, as in the smallest space, tightly compressed; it is that original contradiction, which outwits analyzing and synthesizing reason, but not the generative intuition with the power of its self-certainty.

Trendelenburg believes himself to have demonstrated that space has *at least* three dimensions, through the "extension" of point, line and surface into a body. But the proposition that space can have *no more* than three dimensions is supposed to result from the following consideration: a point inside a body cannot "extend beyond itself" since it is surrounded by neighboring points; this is only possible for the points on the surface of the body. But with that only one more dimension, the third, is added to the two dimensions of the body surface.

Here, indeed, the "power of self-certainty" must help us. Which realm of objects can Trendelenburg possibly be talking about here? One can let a "mathematical body ... come into being" through the "extension" of such forms as points, lines and surfaces. These forms, for their part, must behave like real bodies because such an extension can be prevented by neighboring points. On the other hand this extension is "nothing other than motion itself." And all the amazing contradictions which already lie in the mingling of ideal, mathematical forms with corporeal properties, merely "outwit" the unfortunate mixture of analyzing and synthesizing thought, but are simply not a problem for the mighty intuition!

The modern reader can at least intuitively imagine that these whims are the stuff of which Mickey Mouse films are made. In animation one can extend the (drawn) point into a (drawn) line, and the pen stroke of the animator can flow apart into a surface; and - like poor Mickey Mouse, who is pressed to a two-dimensional being when caught between two blocks, only to spring back a few seconds later, well-preserved, to three dimensions - the colored surface blows itself into a spatial cake.

We have to deal with the practice of drawing here, which joins the Platonic heaven of geometrical ideas to the real world of bodies: a respectable practice, at any rate, which refers the Greek words which were used by the ancients for point and line (*semeion* and *gramme*, literally mark and stroke) to their roots in life. Of course, one has it relatively easy in this context with points, lines and surfaces, but the question of how the practices of drawing and painting arrive at a spatial, three-dimensional cake by painting a surface with a fixed amount of lines remains open - unless one lays it on very thickly.

Trendelenburg himself seems to be uneasy about his deductions, since he closes his proposal with a very resigned remark: "The question is whether, in this intuitive domain, any reason other than that of intuition can be found. In any case, it has not yet succeeded."

We would therefore like to turn our attention to the third and last author, William K. Clifford. In his book, *Seeing and Thinking* (1890)[10] he undertook the most clearly formulated attempt to deduce the three-dimensionality of space from the concept of motion. In the concluding chapter, he explains, under the title "Of Boundaries in General," how one can determine boundaries step-by-step, proceeding from bodies in declining order from surfaces and lines, down to points. In this way, one gains definitions for the fundamental geometrical terminology, in order to then add the concept of continuous movement, from which a triple spatial mobility of the point and, in this sense, the three-dimensionality of space would result. Clifford himself would like this to be understood as a mere explication of our old familiar concepts of space, surface, line, and point.

The beginning, neither explained nor grounded, lies in the claim that "things" - later it is made explicit that bodies are meant - "take up room." To take up (or fill out) room is, so to speak, the pre-understanding which is the basis for everything further and is supposed to suffice as an explanation of "space" (but not of thing or body). Every body generates two spaces in this way: the one which it fills, and the one which it does not fill. The transition from the inside space of one body into the inside space of another body will normally ensue through an additional space (like air). This case is clearly differentiated from the transition of the inside space and the outside space of *one and the same* body. In this case the transition movement does not lead through the inside space of another body, but only through its surface. One thinks of a stone which lies in water, for instance. The surface itself takes up no (corporeal) space. The inside and outside spaces are adjacent. The surface is the outside surface of the body as well as the inside surface of the non-body, or surrounding body. From then on Clifford speaks of regions. The regions (inside and outside space) are divided by the surface.

Clifford stresses that "surface," in this sense, corresponds to Euclid's definition, according to which a surface has length and breadth, but no thickness. So a surface is not like a layer of paint which not only covers a body, but which also takes up space itself.

Now it is a characteristic of the surface of a body that it moves with the body. Every movement of such an imagined surface is continuous, i.e., it assumes an infinite number of intermediate positions between the start and end positions. Clifford warns several times against seeing anything mysterious in the talk about "infinite."

---

10 W. K. Clifford (1890), pp. 127-156

Infinite simply means "without end," which, in the case of movement, means that between every two positions of the surface another can be found, and so on. Clifford clearly represents the Aristotelian view of potential infinity here; just as his entire discussion can be seen as a concise repetition of the Aristotelian continuum theory. (Clifford does not, in fact, mention Aristotle, but says that the mysteries of the continuum and infinity are to be blamed on the fact that one is accustomed to using Latin and Greek words for them; when one translates them into English, however, one sees that the matter is quite clear and simple.)

A further peculiarity of the surface of a body, which moves along with this body, is that after a movement it nevertheless stays back at the starting position, namely, where the body *was*. In this way Clifford can talk about regions or positions independent of whether these are currently taken up by a body; it is sufficient for purposes of illustration that they were taken up earlier. Two regions are called congruent when they were filled by the same body at different times, but only under the condition that the body in question has not changed its shape or size. (What this is supposed to mean, and how one could test this if necessary, is left to the semantic understanding of the reader. The problem of form and magnitude constancy is not considered here.) One sees here how indispensable the concept of motion is, since the traditional geometrical concept of *congruence* is gained from the various positions of a body at various times, and is thus *not a purely spatial* concept.

After this elucidation of the concept of the surface as the body's boundary, Clifford proceeds to the boundary between surfaces. For this purpose he considers a piece of paper with a red spot; analogous to the inside and outside spaces of a body which are divided by a surface, the inside and outside spaces of the red spot are divided by an (imagined) line. So "line" is not a drawn line, no matter how fine the stroke. (A drawn line would, itself, cover a surface and would be bounded by two imagined lines.) It is rather the boundary between two surface-regions which does not take up any "surface-room" itself.

Within the context in which the imagined line divides two surface-regions, and these again divide two solid rooms, it is reasonable to claim that lines "[exist] in space." Clifford ties another comment onto the last which is decisive for his definition of "dimension," as we shall see later. He claims the following: because the line "exists in space," it does not matter *which* surface it divides - it is sufficient that it divides one at all. From this statement Clifford later derives the claim that a line is an *intersection* of two surfaces and not just the boundary of one of them. After all, there is no reason to choose one particular surface from the set of all possible surfaces which bear lines. Clifford obviously forsakes the history of the generation of forms which are being treated here, according to which a line can lie on the surface of a body. We shall come back to this later.

Now a line can be divided again by another line. One takes a second sheet of paper and holds it over the first in such a way that a light-shadow boundary falls over the white-red boundary and intersects it so that the red spot lies partially in shadow. Then the light-shadow boundary "divides" the white-red boundary, and this defines the point. All earlier analogies are repeated: the point takes up no "line-room" and all questions about quantity are meaningless for it. As Euclid said, the point has no parts; it cannot be measured. According to Clifford the point thus has four different properties:

1. It is not only the boundary of a line, but also its intersection - and here are three forms of the intersection which must be differentiated;
2. the point as the intersection of two lines on a surface (the point divides the white and red boundary as well as the boundary of shadow and light);
3. the point as the intersection between a line and a surface which the line penetrates; and finally,
4. the point as an intersection of three surfaces (which can easily be illustrated by thinking of the corner of a room where three different walls (surfaces) come together).

After proceeding from body to point, all of the following concepts are considered in reverse order as the "path" of the previous form in movement. Clifford is, to my knowledge, the only author who first goes through an analytical round of argumentation to gain the geometrical terms "surface," "line" and "point," in order to, proceeding upward from the point, arrive again at the body and corporeal space by using the additional condition of movement. When a point moves, it moves along a line. It seems that what Clifford means here is that the concept of the line as a boundary between two surface-regions is, in fact, already available, but that the *continuous structure* of the line can first be understood through the movement of a point, and not through its property of being able to be the boundary of a surface. (Clifford goes beyond the Aristotelian continuum theory here, where the point is, indeed, indispensable as a boundary for the continuous division of a segment, but where the continuity of the segment is defined independent of motion and time.) "The movement of a point" is exemplified by Clifford through the motion of a body which throws a shadow in such a way that the light-shadow boundary moves along the color boundary. "Moved" is thus terminologically tied to the motion of a body, which is caused by a person. The difficulty which arises with other authors - that it remains unexplained how the predicate "moved," which was introduced in the world of bodies, should be applied to ideal forms like points - is not present here.

These considerations of Clifford's have led to the following outcome:

1. There are *three different spaces*: corporeal, planar and linear space;
2. there are *three different boundaries*: surfaces, lines and points;
3. *four different intersections* are distinguished: surface/surface, surface/line, line/line and three surfaces to each other; finally
4. there are *three different paths of moved boundaries*: the body as the movement of a surface, the surface as the movement of a line and the line as the movement of a point.

After this, Clifford turns to the question of whether space can consist of points, and discusses this in reference to the question of whether a line can consist of points. In the process, Clifford repeats Aristotle's doctrine of the continuum without, however, making any reference to it. According to this older theory, no line segment can be made up of points, because an infinite number of points can lie between any two given points. Lines and points are qualitatively, not quantitatively, different; so it is not as if a sufficient number of points could eventually produce a line. Only point series (as the mathematical equivalent of point movement) generate lines. Accordingly, space does not consist of points, but of a series of points. But while, for a line, a singly infinite amount of points form a series, for a surface, over which a line can continuously move, a doubly infinite amount of points would form a series, and for space, as the path of a moving surface, a series is made up of a triply infinite amount of points.

Thus, the point has one variation of movement along a line, two along a surface, and three in space. To that extent, one could also differentiate the three types of space through *dimensions*: corporeal space has three variations of movement, planar space has two and linear space has one. "Dimension" is then defined as a variation of a point in regard to its movement.

The *triad* of dimensions depends on the bond to the variation of movement of *one* point. If, however, one considers a *pair* of points, then one of them has three primary variations, but the second has, additionally, infinitely many possibilities of distance. So the addition of "size" already raises the number of dimensions to four, namely "three of position and one of size." And then the text breaks off without any warning, with a flowery appeal to take such a discussion, which leads back to the familiar ground, seriously after all.

Hence, Clifford's idea is to first derive geometry's fundamental concepts of the surface, the line and the point from a type of analysis of the body in order to gain with these concepts, and the additional concept of motion, a kind of synthesis about the three-dimensionality of space as the movement possibility of a point. As attractive as this approach is, it is also clear that it has irreparably failed, for the following two reasons:

1. The pleasingly clear definition of the surface of a body as the boundary which must be passed through when one crosses from the inside of a body over to its outside or outer region, is already an *imagined movement of a point*. The definition of surface borrowed from Euclid ("with length and breadth, but without height, or thickness") cannot otherwise be won. In taking an (extended) body - one could think of a small ball hanging from a thread - and dunking it completely in water, and then slowly pulling it back out through the water's surface, one must differentiate between several stages of the passing of the ball through the water's surface, among which, for example, contact with the surface would have to be included. In this way one could not clearly define the surface of the water, but would once again have to distinguish a mathematical surface on the ball, in order to discern the inside and outside spaces of the water quantity as divided by a mathematical surface. (The entire question of whether or not real bodies are, in fact, always so sharply divided from their surroundings, as the assumption of an ideal mathematical closure surface suggests, is ignored here.) The beginning of the Cliffordian approach is circular in both cases, i.e., with the imagined movement of a point and with the passing of a body through a surface, whereby the penetrating body also has a closure surface. This circle cannot be broken when the point itself is supposed to be defined, via surface and line, from the movement of a point passing through a surface.

2. Lines were defined as boundaries between *one and the same* surface. The claim which Clifford made that a line is a *intersection of two surfaces* is a spring that is made without any definitional or logical safety nets. Indeed, even within the structure of the Cliffordian approach there is no plausibility for this, as surfaces are defined as the mathematical closures of bodies. Under these circumstances, what is "intersection of two surfaces" supposed to mean? No passage distinguishes certain *forms* of surfaces, like the plane, so that one might consider labelling an edge as the intersection figure of two flat surfaces, and then consider this to be the dividing line between two regions.

One can follow Clifford when he says that lines have a position in space, but the deficiency just mentioned prevents a sufficient definition of the point. The quadruple definition of the point (as the boundary of a line, or the intersection of line/line, line/surface and three surfaces) proves to be too weak for the following claims about the mobility of a point in space, for the following reasons: the first two definitions, according to which a point is the boundary of a line as well as the intersection of two lines, are, in principle, identical. The boundary character is also explained as the intersection of two lines (with the example of the light-shadow boundary and the red spot on white paper). The third and fourth definitions also collapse into one. The intersection of a line and a surface (in the sense that both share only one point) and the intersection of three surfaces, obviously portray the same state of affairs, when the line is, for its part, defined as the intersection of two surfaces. When the four-fold definition of the point has been reduced to two different definitions, it can be shown why these two are insufficient to make the three spatial movement possibilities of a point and, thereby, the three-dimensionality of space, comprehensible.

The first definition (point as boundary or the intersection of a line) has the mobility of the point along a line covered. The mobility of lines over surfaces is also provided for by the light-shadow example. But concerning the movement of the surface through space, one misses a sufficient definition. The thought which appears in the context of a definition of congruence, that a body can take up different spaces at different times (as the beginning and end positions of motions), must either make reference to another body and a distance between the two which changes with time, for which the necessary definitions are lacking, or return to the idea of absolute space. The second definition of the point (as the intersection of three surfaces) suffers from the fact that a definition for the intersection of two surfaces is lacking. The example which Clifford gives of the corner of a room, in which "three walls intersect at the corner point," would only be an appropriate example if it were also said to what extent two *different* surfaces meet at each edge of the room. To do this, however, one must speak about the form of the surfaces, which Clifford avoids doing.

To be helpful, one could attempt to save the Clifford proposal by assuming the walls of the room to be of different colors, corresponding to the example with the red spot on the white paper, so that three lines which conform to Clifford's definition of the line meet in the corner. In *this* sense, we would have three "intersections" (the edges of the room). But Clifford cannot endorse this interpretation, because then the form which arises from the three surfaces would only be two-dimensional in the sense of the movement potential of the point. This is immediately obvious when one imagines three different colored spots on a sheet of paper, e.g., red, green, and white, meeting in such a way that the three boundaries (one between each two colors) come together. Briefly stated, there would only be one point in which all three paint regions touch. With that, however, we would have a new definition of the point which Clifford had not intended, namely, as the intersection of three lines (which end in the intersection). But here the point would only have a doubly infinite movement potential: the red/green boundary could move along the white boundary, or the red/white boundary could move along the green boundary. But then all the movement possibilities of the green/white boundary along the red boundary have already been realized.

Thus, it has been shown that the lack of a definition of the intersection of *different* surfaces (as opposed to the dividing line between regions of *one and the same* surface) only leads to a two-dimensional space, when "dimension" is understood as the potential of a point for continuous movement - even if this surface were to have the form of the corner of a cube.

Now one can repair the deficiency which arises from not speaking about the spatial *form* of a body or its surface: one imagines a body *B* with a color spot (region *R*) on it; a second body, *B*', (made from some plastic material) is placed over *B* in such a way that *R* is precisely covered by *B*', without overlap. In this way a region *R*' is bounded on *B*' which, according to definition, coincides with *R*. The boundary of *R*' on *B*' is, however, (every bit as much in accordance with definition) also the boundary of the remaining surface *S*' of *B*', so that now a new state of affairs has come about. In the boundary of *R* on *B*, the remaining surfaces *S* and *S*' of *B* and *B*' also come together. For this reason, one says, per (new) definition, that different surfaces (because they lie on different bodies) intersect here.

This is, of course, a liberal supplement to Clifford's text, but it is logically consistent. This supplement contains something principally new because it must make use of the fitting of two bodies to one another, and thus would have, as its premise, the physical equivalent of the geometrical concept of the congruence of two surfaces (with different spatial orientations!). In this way something principally different would come into play, to wit, the concept of congruence with regard to the orientation of surfaces. Clifford's proposal could be corrected of this deficiency of not having defined the intersection between two surfaces through such a "form-theoretical" addition, namely, that concerning the fitting of two regions on bodies. But this does not correct the first mistake mentioned above, that the surface of a body defined as a mathematical, infinitely thin closure surface results only when the continuous movement of a point through a body is already terminologically available. However, this movement of a point is, according to Clifford, only supposed to be derived *after* the analysis of the relations of a body. His approach is, therefore, a mere *petitio principii*.

SUMMARY

The entire second section of this chapter was devoted to the question of whether the three-dimensionality of space can be traced back to the addition of motion, to other explanatory models from geometry, or a more-or-less theoretically constructed description of bodies. The answer is surprising. As long as the argumentation is purely mathematical - where "motion" is only a useful metaphor for certain constructions, mappings or rules of operation, detached from any reference to the world of bodies - a restriction to three-dimensionality is not wished for, and is therefore not maintained. But to the extent that a reference to motion in any given text is a reference to the motion of real bodies, it is the properties of those bodies which have three-dimensionality as a consequence, and not the properties of motion.

In order to make this clear, let us briefly review the individual approaches once again, without holding strictly to the sequence in the text above. A systematic presentation is better here.

All the motions that have been considered at all are so-called rigid motions. If figures are referred to, then these should be congruent at the beginning and end of the motion. This is assured through the construction of a parallel shift of all points along the same distance, or through a corresponding rotation construction. In contrast, if real bodies are referred to, then these should behave as if they were bounded by geometrical figures, in which case the same rules apply. In other words, the "concept of motion" is always considered to go back in definition to construction rules. To the extent that geometry is not understood as merely formal theory, i.e., when "construction" means actual interaction with bodies, like the construction of a figure on the drawing board with the aid of a compass and ruler, the *reference to the world of bodies is already provided by the drawing instruments*. These must actually be moved while drawing - by a person, of course. But here, too, one will not permit the drawing plane, compass and ruler to be made of soft rubber; they must be sufficiently hard and constant in form. If this form constancy is to be raised to the rank of a scientifically controllable quality, and not simply to be assumed naively or in terms of ordinary language, then it must be described or defined from the very start. These definitions do not appear in the approaches which have just been discussed. The form constancy of moved bodies, surfaces and lines is, in fact, simply assumed by Kant and Clifford, as well as Natorp and Trendelenburg. None of them sees the inherent problems of realization and definition. (A contrary systematical claim is that, in solving this problem, three-dimensionality is already achieved purely spatially, that is, without motion theory. The second part of this book will examine the validation for this claim.)

The problem of incongruent counterparts, i.e., of forms which are mirror-congruent without also being congruent, results from several conceptual obscurities or omissions. As a purely geometrical problem, the definition of orientation - as that property which characterizes the existence of figures which are mirror-congruent, but not congruent - is possible for any given dimension number, and consequently does not provide an argument for the number three. If orientation in a geometrical sense is assumed, then the question about the convertibility of real bodies into their mirror-congruent counterpart proceeds either from a nonsensical claim or from a misleading metaphor. To convert a body into its mirror image - in the literal sense of "mirror image" - is quite simply absurd. In the metaphorical sense, two equal (in perhaps every property but orientation) bodies are present - and then talk of "conversion" is as pointless as the conversion of two billiard balls which are equal, but of different colors, into one another.

It follows from the definition of orientation that an oriented body cannot be converted, by motion, into an otherwise equal body of the opposite orientation. The "analogy argument," according to which this conversion is supposed to be possible for bodies of lower dimensions by motion through a space of a higher dimension, is untenable. All bodies, regardless of whether they have a particular orientation or not, are three-dimensional and behave in the same manner, even if they can be misunderstood as being one- or two-dimensional through the omission of certain spatial aspects in their descriptions. In other words, when the argument of congruent counterparts or the analogy argument is completely formulated, the three-dimensionality of the bodies is always involved. It is never their motion (or the impossibility of motion), which delivers the argument for the three-dimensionality of space.

In the end, imagined motions - the rotation of an oriented line segment into a plane, and the plane into space (Natorp); the extension of the point into a line, the line into the surface and the surface into space (Trendelenburg); or, finally, the continuous movement of geometrical forms, which are defined as boundaries (points) of boundaries (lines) of boundaries (surfaces) of bodies, along lines and surfaces, and at an angle to these surfaces (Clifford) - all gain their intuitive certainty through a definitional access which begins with bodies and distinguishes the triad of the forms point, line, and surface, as properties which define, or bound real bodies, which assure their existence, or which deliver their Euclidean description, *before any discussion of motions*.

In principle, all the approaches discussed here are merely failed attempts to solve the unfortunate definitional problem of the geometrical basic concepts of point, line and surface. It is unfortunate for several reasons: Euclid laid a false trail for the following two-and-a-half thousand years with the sequence he chose for definitions. He proceeded in an ascending order from point to body, instead of the descending order which naturally occurs when cutting apart bodies (which were present before geometry). Then, to be consistent, he defined the line before the straight line and the surface before the plane, even though access to geometry only succeeds with distinguished, recognizable forms like (first) the plane and (then) the straight line as its intersection. Finally, he could not make clear the connection between ideal geometry and the properties of the world of bodies, which we are accustomed to describing with the same words. Indeed, he mixed words for real, drawn figures and the ideal properties of these. For surface and body he did not distinguish at all between the real and ideal realm, but, in fact, used one and the same word in both cases.

All of the approaches spoken of here share the goal found in this tradition. Each wants to extract the properties of space, of the body or its motion *from the object itself*, as it were, instead of taking into account the *active involvement of the person*, who, with his actions of moving or shaping bodies, first brings about the state of affairs which alone can be the object of this debate. In any case, whether neo-Kantian transcendental or analytic-empirical, they are all "objectivistic," i.e., seeking in the object something which has its roots in the acting, producing subject, in his intentions, and in the success of his linguistic and manual practices.

CHAPTER THREE

Argument for Three-Dimensionality from Laws of Force

PRELIMINARY REMARKS

Principally new approaches come into view with the attempt to explain the three-dimensionality of space with laws of force. These approaches are science-analytical arguments in connection with empirical theories. To what extent is this type of argument "principally new?"

It is understood that talk about the three-dimensionality of space is, of course, talk concerning space, i.e., of spatial distinctions, making use of predominantly geometrical words, even in ordinary language. The question about the triad of spatial dimensions is nothing more than the question of how this triad is to be extracted from the form, magnitude or position of a body, or perhaps from the geometrical differentiations of surface, line and point (or their forms, like flat, straight, regular polygon, etc.). Even the extension of the considerations to motion (of bodies, or of geometrical figures which describe them spatially) is not tied together with an extension of the spatial vocabulary. The question about three-dimensionality, as posed, has already fixed the realm of the objects which are referred to. One could also say that it is an analytical triviality that the question about the three-dimensionality of space is raised and discussed within the framework of spatial theories.

The extension of the considered realm of objects to include forces which are at work between two bodies, however, is not an analytical triviality - one could have just as well have chosen something else as the extension. Other ones have, indeed, been chosen historically, as the investigations of the causalistic[11] and perception-theoretical approaches in the next two sections will substantiate. In other words, there must be a good reason to adopt forces, more specifically physical laws of force, as the focus of argumentation for the triad of spatial dimensions. Thus, while one continues with the subject of the

11 Translator's note: I am aware that "causalistic" sounds awful in English. The reason for adopting such an awkward word is to distinguish it from "causal" - which refers to relations of cause and effect. "Causalistic" refers to ways of talking or types of description in which causal relationships play a central role. The same reason applies for words like "mentalistic," "formalistic," "passivistic," "naturalistic," etc.

posed question of geometry, or extra-theoretical spatial statements about bodies, a dynamic explanation of the three-dimensionality of space cannot be considered until a physical dynamic is developed and *at least one law of force* is formulated and sufficiently accepted. Therefore, a dynamic argument is, *per se*, a science-analytical one. This means that a particular scientific theory is the historical and systematical prerequisite of such an argument, without which it would not be possible, much less meaningful.

We shall discuss two approaches in this chapter, namely, the proposal that Kant submits in his "Thoughts on the True Estimation of Living Forces," and Ehrenfest's thoughts on the role of the three-dimensionality of space in the fundamental laws of physics (1917), which were published in 1920 in the *Annals of Physics*.

To the best of our knowledge, Kant was the first who believed to have found an argument for the three-dimensionality of space from the reciprocal dependence of the gravitational force between two bodies on the square of their distance. And Ehrenfest was the first to demonstrate the type of modern analytical consideration that draws conclusions from the acceptance of certain (empirical!) theorems about the premises of these very theorems, in order to then interpret these conclusions as statements that represent an empirical knowledge of the world.

This is also the reason why this analytical attempt to provide a basis for three-dimensionality is viewed critically here: is it true that the principle difference between this argument and the purely spatial approaches opens up a first access to empirical reasons for three-dimensionality, reasons which especially rely upon physical experience? And, to the extent that this is true, is it also true that, because of the historical and systematic dependence of this argument on the formulation and recognition of certain physical theories, no reason to accept the three-dimensionality of space would exist if these theories were disproven? Evidently, the *epistemological status* of the thesis begins to play a role for the first time, in contrast to all previous considerations; in other words, the point in question is whether this thesis formulates an *empirical fact*. Or is it, in the end, merely a *conventional stipulation* of definition? (Or can it be traced back to one?)

From the start, it can be seen that Kant's and Ehrenfest's attempts, when successful, have far-reaching consequences. The (until now) unfounded belief in the three-dimensionality of space, shared by laymen as well as physicists, would be solely supported by empirical, physical theories, and would remain tied to the fate of such theories; for example, in the case of an empirical revision of these theories three-dimensionality would itself be open to revision. This would raise a series of utterly unsolvable problems, not the least of which is the intuitive certainty of the layman that space is, necessarily, three-dimensional, and depends neither on the arbitrariness of definitions, nor on the revisability of experience.

Of course, it will not be enough to determine whether the two authors, Kant and Ehrenfest (who represent here, as prototypes, a large number of similar attempts), were successful in deriving a proposition about the three dimensions of space from empirical propositions of physics. Even if this were the case, it would in no way insure that this proposition is, for its part, an empirical one. Even valid empirical propositions contain sub-statements which do not have to be based on experience, but are simply contained in the empirical proposition. To give an example of this, in antiquity five planets were known from the observation of lemniscates; from the observation of the heavens in daily and yearly rhythms, it was realized that the earth was a planet of the sun. After this, the empirical proposition that the sun has six planets was valid, but the arithmetical proposition contained therein $(5+1=6)$ is surely not an empirical proposition because of this. It will have to be tested in the case of Kant and Ehrenfest whether, in spite of their beginning with an empirical proposition of physics, one of its non-empirical premises was brought to light through the analysis; and if this is the case, whether the entire consideration does not simply proceed from a *petitio principii*, as was already the case with the purely spatial approaches.

KANT AND THE TRUE ESTIMATION OF LIVING FORCES

The title of the Kant text that we are referring to here, already informs us about the context in which the question of three-dimensionality is raised. It is a critical dispute with Leibniz about the difference between *vis viva* and *vis mortua* (in modern terminology, between kinetic and potential energy), as well as an improvement proposal of Kant's. One would not suppose from the title of the inquiry that it would be a discussion of a spatial problem.

After Kant linked the definitional unity of a world with the condition that its things find themselves in a "real connection" in §8, §9 begins with the assertion that "It is easy to prove that there would be no space and no extension if substances had no force to act outside themselves." When "force," "extension" and, as in the next sentence, "order" are evoked, only repulsion can be meant, which, as impenetrability, prevents two bodies from being in the same place at the same time. Thus, without the effect of force of bodies on one another, there is no space. According to Kant it is more difficult, in contrast, to see how "the number of spatial dimensions follows from the law according to which this force acts outside itself."

Introduced by the indication that he finds merely a circular argument to the point in Leibniz's *Theodicee*, Kant presents a short geometrical consideration which is based upon the impossibility of intuitively imagining, geometrically, a fourth "power" (following length, surface and volume). In any case, Kant himself does not attribute much soundness to this argument. He expects instead that the sounder argument lies in "a certain other necessity, which I am not yet able to explain." Therefore it is, according to §10, "probable that the triple dimension of space stems from the law according to which forces act upon each other's substances."

A law which is valid for all bodies exhibits itself in the "type" of interaction between bodies, and also determines the "type of law ... according to which substances seek, by virtue of their essential forces, to combine themselves." Without giving a single deduction or further reason for this, he goes on, "that the substances of the existing world... have essential forces of the type that in combination with each other, they radiate their forces in a double inverse ratio to their distance," which is why "the whole," that is the set of all bodies that interact with each other, "has the property of triple dimension."

This passage is normally understood in such a way that the square of distance between two bodies, which is inversely proportional to the Newtonian gravitational force, provides the compelling evidence for the three-dimensionality of space. The amount of force which affects a body (and, thereby, the amount which radiates from it) goes through a spherical surface which surrounds the body. The spherical surface is determined by the square of the radius, i.e., when the radius is doubled, the surface through which the same amount of force goes is quadrupled. One can supplement the text with the explanatory observation that if the effect were simply inversely proportional to the distance between two interacting bodies, one would have a ratio as is found in a circle, where the doubling of the radius only leads to the doubling of the circumference - and if this were case with the law of gravity, we would have to suppose a two-dimensional world as a consequence.

Kant adds that "this law is arbitrary, and God could have just as well chosen another," which would then have another number of dimensions as a consequence. It would be "the highest geometry" that a finite mind could undertake to treat all possible types of space with various dimension numbers in a scientific manner. The boundaries of our intuitive powers of imagination provide no objection against this possibility, and arise rather from the structure of our perception and our human possibilities for action, which also lose their intensity with the square of the distance.

Going back to the question of §8, about whether a number of worlds can exist independent of one another, Kant stresses once again now in §11 the arbitrariness of three-dimensionality, which is most certainly not the arbitrariness of the perceiving or defining person, but that of the divine creator of the world. Since Kant believes he has already proven the possibility of the existence of several worlds, he now derives (from a reverse inference which is not exactly logically flawless) the assumption that all worlds would be dynamic, that is, would be connected by the law of gravity, if only three-dimensional space were possible. It is not probable that many worlds actually exist, "unless many types of space... are possible," that is, spaces of different dimensions.

We do not want to discuss these wild speculations which even Kant himself characterized cautiously as thoughts concerning "a sketch of a reflection which I reserve for myself." What does need to be analyzed, on the other hand, concerns the following two questions: 1) Can one really conclusively argue for three-dimensionality from the form of the Newtonian law of gravity, that is, from the inverse proportionality of the gravitational force between two bodies based on the square of their distance to one another, and 2) does this argument provide an empirical reason, or does it only bring a definitional (or other) premise of the square law to light?

To go into the first question, it is a well-known and trivial geometrical fact that a point or central projection of a surface leads to a "square sentence," i.e., that a doubling of the distance from the projection point leads to a quadrupling of the surface. This square sentence is a sentence about a three-dimensional state of affairs in normal understanding, since, for its formulation, one needs on one hand, a surface (two dimensions) and, on the other hand, a direction of projection that does not lie on this surface as a third dimension. And this geometrical state of affairs corresponds to a well-known physical one: when one illuminates an area *A* with a (best possible) point-shaped light source from a distance *d*, this area throws a shadow at distance *2d* of the magnitude *4A*, and *A*, brought to a distance of *2d*, would only be illuminated with a fourth of the brightness compared to the brightness at distance *d* (*A* and parallel screen). The geometrical state of affairs of point projection is an idealization of a known physical state of affairs, at any time technically producible. (Since G. H. Lambert, temporally before Kant, had already defined the equality of illumination strength with his grease spot photometer - with equally strong illumination of a sheet of paper from both sides, a grease spot on the paper becomes invisible - this known diminution of the illumination strength could have already provided an indication for the acceptance of the three-dimensionality of space. Hence, besides the Newtonian law of gravity there is another empirical state of affairs, that was known then, which could have been called upon as an argument for the three-dimensionality of space.)

The illustration of the square sentence by a shadow-throwing area *A* shows that use is made of the rectilinearity of the light rays, as well as the assumption that it is exactly the same amount of light rays that strike the area *A* at a distance of *d*, and the area *A4* at a distance of *2d*. This is not a foregone conclusion as regards factual, physical proceedings. The air streaming out of a fine nozzle aimed at *A* "behaves" differently and tends to come together again into an airstream behind the obstacle *A*, or forms whirlwinds, but in no event does it leave a "wind-shadow" of *4A* at a distance of *2d*. In transition from optical to dynamic gravitation-theoretical examples, one must also suppose that the "light rays" that one can in any case still picture as very small, focused bundles of light, correspond to something like rays of force or lines of force, which propagate rectilinearly. There is no indication whatever that Kant had an conception of lines of force or of force fields. Historians of physics accredit this conceptual model to Michael Faraday (1791-1867). Since Kant does not say to what extent he shares the conclusion that the dimension number of space must always be one higher than the exponent of the distance with which an effect of force diminishes, nothing more can be said about the silent assumption that Kant actually made. One can only state the logical and definitional conditions which actually make such a connection possible.

But with that, the second of the questions posed above is already answered: ignoring the experiential character of Newton's law of gravity (in any case, in a mechanics in which the definitional problem of the mass is solved independently of gravitational forces[12]) the connection between force and distance is solely the expression of the geometrical state of affairs of a central projection and is thus not empirical. The connection of the square sentence is valid for geometrical reasons every time a rectilinear propagation is assumed, the source of which is imagined to be point-shaped, with no preferred direction of propagation - even this is a strong supplementary assumption in every single case of application.

Suppose that the experimental control of the diminution of the gravitational force between two bodies with the square of their distance is successfully achieved, even in an otherwise gravity-free region for every given spatial direction, this law requires the three-dimensionality of space neither as hypothesis, nor as logical consequence. (One must, of course, be able to speak about the straight connection between the two bodies, as well as about the

---

12 As occurs in the essay by B. Thüring (1967).

determination of the independence of force from direction.) The conclusion that Kant draws however, that three-dimensionality follows from the *square* of the distance, that is, the exponent two, must unavoidably make use of the geometrical state of affairs which is already contained in the sentence about central projection. Hence, it is *not the law of gravity* itself, but the *conclusion of three-dimensionality from the law of gravity* is the passage in which the geometrical state of affairs that the area of a spherical surface depends on the square of its radius comes into play. The spherical surface as the envelope of a body, however, is necessary as the condition for all the forces affecting this body to be continuously propagated and registered.

In other words, the law of gravity, to be tested by measurement along a straight line, as a diminution of the gravitational force between two bodies squared with the increase of their distance from one another, must first be provided with an additional assumption of isotrope (i.e., equal in all directions) propagation, so that the image of the sphere which encloses the gravitating body and through which the gravitational force must completely penetrate can then be applied, in order to apply the square sentence. And with that it has been shown that it is not the law of gravity itself, that is, not the empirical state of affairs of the dependence between force and distance, which establishes the conception of three-dimensionality.

The first historical attempt to draw upon an empirical physical law to establish the three-dimensionality of space, proves to be a *petitio principii* - which should not be too surprising, because an actually undertaken measured control of statements of physics must necessarily entail dealing with bodies and make use of a technical mastery of their spatial properties such as form, magnitude and position. And for this, no one has ever used anything other than a three-dimensional geometrical conceptual system. We shall also run up against this simple fact (overseen within empirical prejudices and self-appraisals of physics again and again) in the next, modern variant of the Kantian argument.

EHRENFEST AND THE STABILITY OF ORBITS

A lecture which Paul Ehrenfest gave in 1917 appeared in the *Annals of Physics* of 1920 under the title "What Role Does the Three-Dimensionality of Space Play in the Fundamental Laws of Physics?" The title does not, at first, lead one to expect that its thesis is that space *is* three-dimensional and that one can perceive this in the fundamental laws of physics.

Ehrenfest begins his presentation with an extremely cautious preliminary remark, according to which the question "why?" in reference to the three spatial dimensions is "perhaps meaningless," but "in any case ... calls for critique,"[13] and he adds the hope that a "reasonable" question could be "constructed" later to the answers which have already been found.

Benevolently interpreted, there is a reference here to a strict analytical procedure: acknowledged theorems of physics are analyzed in the form that, proceeding from the normal three-dimensional form, consequences are discussed, when a two-, four- or higher-dimensional version is *formally* considered. In other words, an empirical proposition, which (trivially) has an experimental confirmation "in the three-dimensional," i.e., with instruments, and so with spatially manipulated bodies, is with the normal means of mathematics, given a more general, namely, an n-dimensional, formulation. This is only the application of another way of writing (mathematically completely unproblematic). But then the new way of writing is tied to the supposition that the proposition in question has also changed its content with the change in the way of writing, and is, thus, an empirical proposition "in other [other-dimensional] spaces." It becomes clear here that this procedure throws the conditions of validity of the discussed fundamental laws overboard, which destroys the reliability of the drawn conclusions.

Ehrenfest proceeds by seeking a "singular" property of *S3* in physical laws, which is not given for *S2* and *S4* (or higher-dimensional spaces). The result (formally unobjectionable) is that only in *S3* do circular orbits remain stable against small disturbances in closed orbits between two (finite) radii. This coincides with the observation of planetary orbits and their explanation through the Newtonian law of gravity. Ehrenfest expressly mentions that the approach to the law of gravity accepts the following premises: that the force which affects a body is "directed at the center and is only a function of *r*," and that "the theorem of Gauß about the streams of lines of force should apply even in *Sn*." Without going into its mathematical form, it can be seen that Ehrenfest holds the same premises (here obviously based upon equation) which were already decisive for the discussion of the Kantian proposal above; the entire force which affects a body goes through a sphere which surrounds the body, and this is only dependent on the distance from a counter-body.

Therefore, the same objections are raised again: the empirical validity of the basic connection between distance and force can only be achieved and controlled within the frame of three-dimensional laboratory physics, just as the actual planetary motion is, indeed, only

---

13 P. Ehrenfest (1920), p. 40

accessible as an astronomical observation result in *S3*. (Or is there any other possibility for astronomers to discern the planetary motions as stable, closed orbits when limited to one observational plane in which the observer would himself have to lie, that is, when limited to a "two-dimensional observational space"?)

Thus, the fundamental laws discussed by Ehrenfest are only meaningful propositions relative to the research practices of physicists, described as three-dimensional; and, *applied to these*, the quoted Gaussian theorem is only congruent with the square theorem again, i.e., one also invests the premise of three-dimensionality already in the mathematics applied here, even if it is much better hidden here than in Kant, since the Gaussian theorem is, indeed, "generally" formulated, i.e., formulated for n-dimensional spaces.

At any rate, Ehrenfest himself called the question about the grounds for three-dimensionality of space "fascinating," and adds in a "closing remark" that there are a number of counter-examples against the singularity of *S3* - above all for the numbers 2 and 4, by which the latter can be interpreted as a space-time continuum (within the frame of Maxwellian electrodynamics and relativity physics). This closing remark, together with the cautionary measures quoted above, make Ehrenfest's result questionable. In the end, a clear thesis about three-dimensionality is still missing. A distinction of the triad is discovered in propositions which, in spite of their character as "fundamental laws of physics," nevertheless are formulated by people - but whether this distinction was already unconsciously inserted into these propositions, and if so, where, remains unasked. Ehrenfest's attempt remains an analytical game with formulas, without epistemological worth.

THREE-DIMENSIONALITY AND MODERN PHYSICS

Kant and Ehrenfest were cited as examples of the attempt to explain the three-dimensionality of space from empirical physics. Kant, because he was the first, to our knowledge, to take a law of force, specifically the dependence of the gravitational force from the square of the distance between gravitating bodies, as evidence for the three-dimensionality of space. And Ehrenfest because he was the first, to our knowledge, who presented this argument within the formal frame of modern physical theories. In the physics of this century there have been many attempts to draw upon the concept of dimension (in the meantime, made more precise by mathematics) for the analysis of physical theories, and to derive "empirical" statements about "space" from this. Max Jammer values the attempt of Hermann Weyl most highly. This attempt shows that an effect magnitude as integral invariant only exists in a (3 + 1)-dimensional space-time manifold, on which the Maxwellian electrodynamic is based.

We do not want to go into this and other attempts here. Instead we will generally discuss what epistemological status an argument for the three-dimensionality of space can have when analytically derived from physical theories. Attempts like those of Ehrenfest and Weyl can be benevolently interpreted as beginning with acknowledged, that is, *empirically confirmed*, physical theories, and discussing their formal properties (like the presence of universal invariants), in order to track down *silent premises, of which the theories' founders were not explicitly aware*. Seen in this way, the contributions of physics to the dimension problem of space (more general ways of speaking, according to which any perimeter of a physical theorem is labelled as "dimension," are not considered here, of course) cannot be understood as general questions - as free from assumptions as possible - about the structure of our empirical space. But they are highly meaningful as special questions about which premises about space lie at the foundation of modern physical theory. The answers - assuming logical and terminological correctness - are *relatively* valid, namely, relative to the validity of the analyzed theories. If this is absent, then the analytic implications are also open to revision, and in need of it.

We now assume that analyses, like those by Ehrenfest or Weyl, of prominent theories like classical Newtonian mechanics (including the law of gravity) or relativity electrodynamics, result in the distinction of three-dimensionality above other dimension numbers. Stated more simply, we assume that the physical theories acknowledged today, to the extent that they exhibit corresponding properties, "are valid for three-dimensional space." What kind of knowledge is this? In particular, has physics then proven something that otherwise would not be known, or known, but with a lesser reliability? In other words, aside from a (very meaningful) knowledge about the structure of certain physical theories, is anything about the structure of space also discovered?

The answer is no. And this result can even be principally supported for the results of all physical theory analyses. Regardless of which silent assumptions are responsible for particular theories, they all have the fundamental understanding in common, that all physical theories must be submitted to experimental control in order to be regarded as "physically valid." One knows from the modern theory of science that the connection between experiment and theory is much more complicated (and also looser) than it was earlier thought to be. The idea of the experimental verifiability of physical theories finds itself in permanent retreat: from the all-decisive "*experimentum crucis*," inductive conclusions, and the more-or-less tried and true theory concerning falsifying experiments, to resigned decisionism, where only the bald decision of the experts is supposed to produce the connection between experiment and theory.

The professional physicists, however, have fortunately not been influenced by the failure of the theoreticians of science to properly characterize this connection. Whatever this connection might be in a logical or methodological sense, physics lays claim to "validity in reality," which means that its statements are supposed to be valid for experiential reality, and are supposed to be based upon experience. This is derived from observations, measurements and experiments, in whatever single steps are necessary.

This claim, and its validation through the actual research practices of the physicist, applies to all of the physical theories that are available for dimension-theoretical analysis. It gives them a common technical foundation, which seems perfectly obvious, when three-dimensionality is exhibited analytically in proven theories, too. In particular, the experimental and measuring control of physical statements is a practice which we exercise and grasp three-dimensionally, *prior to - and independent of* - physical theory. The craftsmen and the industries which specialize in producing laboratory supplies, would doubtless be a little perplexed if two- or four-dimensional instruments were demanded of them. Instrument construction and production, as well as their application by the laboratory physicist, always primarily concern the technical (and conceptual) mastery of corporeal forms. Components of instruments are poured, cut, lathed or otherwise produced through the working of insufficiently (or not yet sufficiently) formed pieces of material. The magnitudes of the components must fit one another and their position must correspond to their function. The form, magnitude and position of the components, however, are described (and mastered) with a geometrical vocabulary, which does not principally differ from that of the architect, the carpenter or the mason. In short, the states of affairs that are treated in physical theories, including their temporal sequences and causal relationships, are produced artificially in a terminologically and technically three-dimensional world. So in the end, it is not very surprising when this three-dimensionality in theories is, on occasion, analytically brought to light.

One can discern how little knowledge there is to be gained from dimension-theoretical analyses like those of Ehrenfest and Weyl. On one hand, because of the logical sequential relation between physical theory and analytically-inferred three-dimensionality, it follows that the latter is no longer valid when the former is held to be in need of revision for empirical reasons. On the other hand, only a very specialized argument for three-dimensionality would be lost. The consequence of this, however, is not that space would then have to be attributed another dimension number. The physicists themselves, who would naturally hold onto their prior laboratory practice of using three-dimensional instruments, would be the last to accept such a consequence.

SUMMARY

Attempts to prove the three-dimensionality of space with laws of force or, more generally, with physical theories, bring something principally new into play. As opposed to purely spatial or kinematic considerations they have premises that are based on scientific experience, and so rely on measurements and experiments.

The "pre-critical" Kant begins it all by taking a "natural law," exemplified by the Newtonian law of gravity, as evidence for the three-dimensionality of space. In this way, he becomes the predecessor of "analytical" considerations, even though he delivers no extensive proof for the connection between the law of gravity and three-dimensionality.

Such considerations were first outlined in our century, beginning with Paul Ehrenfest. One can understand them as questions about which silent premises regarding the structure of space and its dimension number are assumed in acknowledged, empirical theories of physics. Analyses of this kind bring a knowledge about physical theories to light, but are not to be interpreted as knowledge about space. Their results remain relative to the agreement with the respective physical theories.

The fact that some theories distinguish the triad of the dimensions of space is not at all surprising. The state of affairs of which they speak, more precisely their experimental presentation, is the technical interaction with bodies in the manufacture and application of instruments for laboratory physics. As such, it forms the technical foundation of the analytic theories. Without such a foundation, these would have neither an object, nor validity. The only known language appropriate for interaction with instruments is that of a three-dimensional geometry in which forms, magnitudes and positions of bodies are described.

To that extent, these analyses also deliver no empirical knowledge about space, contrary to the understanding current in empirical physics, because the thesis of the three-dimensionality of space, even if it is logically derived from empirically valid theories of physics, is not empirically valid. Rather, as the technical practice of laboratory physics, it is jointly responsible for the object constitution of the respective physical theories, and accomplishes the verification of the claims of validity which the physicists make for their theories, in the form of instrumental means for the gain of empirical data.

CHAPTER FOUR

Causalistic Explanations and Three-Dimensionality

PRELIMINARY REMARKS

It is self-evident that the three-dimensionality of space must, in the end, be grasped as a description of *spatial* states of affairs (*Sachverhalte*).[14] It is, in any case, open as to which "things" (*Sachen*) "behave" (*verhalten*) spatially in three-dimensions, and which reasons suggest (or even compel) their *three*-dimensional description. The purely spatial approaches seek reasons in geometry or in the description of bodies, their form, magnitude or position. The supplementation of motion (of bodies or of geometrical forms) assumes that the triad cannot be proven purely spatially. Finally, in the approach which deduces three-dimensionality from physical laws, the addition of the effects of force is taken as an argument for three-dimensionality for the first time. These are the three groups of approaches that have been discussed up to now.

There are attempts by authors in this century which differ from those above in that they try to confirm the conventional choice of three-dimensional descriptive means within the frame of Analytical Philosophy and theory of science from considerations which concern the possibility of cause-effect relations. The simplified thesis states that only a "three-dimensional language" - perhaps with the addition of an economy postulate - allows the normal and tested description of the causal connections between circumstances or events.

Two prototypic approaches will be introduced here: in essence, Honor Brotman's essay "Could Space Be Three-Dimensional?"[15] concerns the area of pre-scientific and extra-scientific ordinary language, while Rudolf Carnap's article "Three-Dimensionality of Space and Causality. An Investigation of the Logical Connection Between Two Fictions"[16] treats scientific causal explanations in physics.

---

14 Translator's Note: The German word "*Sachverhalt*" is a compound made up of the German words for "thing" (*Sache*) and "behavior" (*Verhalten*). Hence, a "*Sachverhalt*" is a description of the way things behave. For lack of a better alternative, I have translated this as "state of affairs."

15 H. Brotman (1969)

16 R. Carnap (1924)

THREE-DIMENSIONALITY AS THE CONDITION FOR EVERYDAY CAUSAL EXPERIENCE

The words "cause," "effect," or "causality" do not occur in Brotman's essay. However, treating this text as "causalistic" is justified by the decisive argument that Brotman presents against a two or four-dimensional ordinary language, in favor of a three-dimensional one. (Of course, it is only an abbreviated way of speaking here to call a language n-dimensional; what is meant is that in it words and phrases are commonly used which demonstrate the distinction of a certain dimension number.) This argument states that a two-dimensional language is too poor, and a four or higher-dimensional language is too rich, to describe what an observer sees in motion (his own, or that of another body). Functioning as the criterion (not further explained) for this, is the fact that otherwise (that is, in a non-three-dimensional case) additional causal hypotheses would be necessary for the choices made in establishing language conventions: e.g. in a "two-dimensional world" concerning the disappearance and reappearance of bodies which are covered by another body for a time, and are thus invisible to the moving observer.

The question about the four-dimensionality of space can mean, according to Brotman, at least three things:

1. Is there in addition to length, breadth and height, a fourth something of this kind?
2. Can more than three straight rods come together (pairwise) at right angles to each other?
3. Can we use four coordinates to determine position (of bodies)?

Brotman presents a consideration, in a rather sloppy terminological and argumentative form, in which all three interpretations of the question are brought into connection with each other:

To begin with he simulates a film - simplified to the two-dimensional case (can more than two straight lines stand orthogonal to each other in a plane?) - in which the actors succeed in producing such an arrangement. It is assumed that "orthogonal" is defined for pairs of straight rods by the condition that any two right angles are equal (congruent) and would, by lying next to each another, result in a straight line. The definition of "straight" is assumed to be ostensive, or based on examples, i.e., a rod is straight when it "looks straight." The word "straight" thus has no other definitional status than the color word "red." One has practiced the correct usage on examples and counter-examples. A mathematical geometry does not come into play here.

For Brotman, the given illustration suffices for the assertion that it is an *empirical* question, whether or not more than two rods can stand perpendicular to one another in a plane. And empirical states of affairs can also be principally different, because "we can imagine something different."[17] The performance (presented in the simulated film) is supposed to demonstrate that in the execution of the proposed angle comparisons, appropriate distortions occur so that every two of the three rods are always perpendicular during comparison to one another. The point of the argument lies in the fact that the "distortion" is "not merely a question of sensual perception"[18] but that a choice, a stipulation, is demanded. Just which stipulation is needed is discussed in view of the two other interpretations of the initial question, that is, questions 1 and 3, in the following manner.

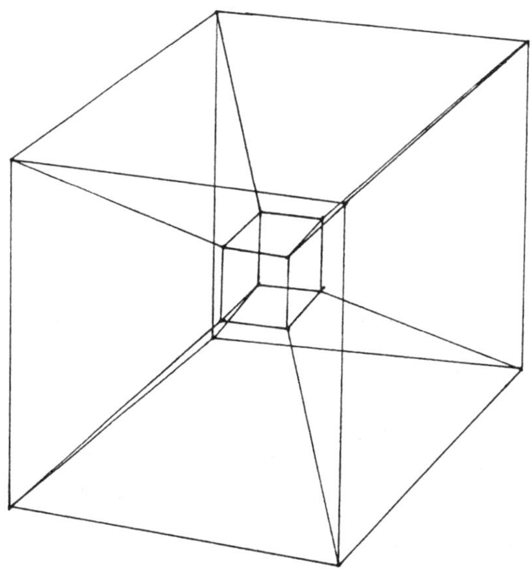

Figure 3

Once one has had (a fictive) access to the state of affairs in which four rods, (pairwise) perpendicular, could come together in a point then one could - now back in a normal empirical space - also consider the commonly known construction of a four dimensional "cube." (Cf. Fig. 3.) Brotman compares this to an unusual (two-dimensional)

17 H. Brotman (1969), p. 52
18 Ibid, p. 54

picture of the normal three-dimensional cube (Cf. Fig. 4), which can be understood such that one is looking perpendicularly and centrally at a surface of the cube and, at the same time, is looking into the cube (which is outlined as an open, cube-formed box), in such a way that the back surface of the cube is the small square. The picture of the four-dimensional "cube" is then interpreted analogously. (Brotman: "The analogies are clear."[19])

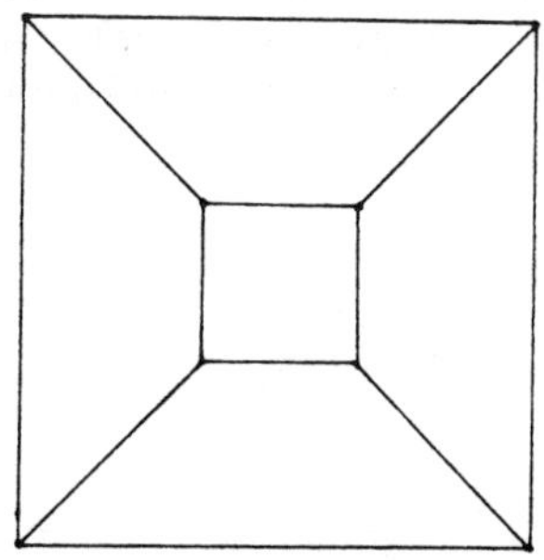

Figure 4

Together with the position argument, it is asserted here that we can "imagine" a fourth something-of-the-kind like length, breadth and height. With a three-dimensional cube, the small square which represents the back surface is smaller than the front one because it is farther away from the observer. By analogy, with the four-dimensional cube, the inner, smaller (three-dimensional) cube can now be interpreted as a cube of the same size as the cube around it, only farther away from the observer than the (three-dimensional) cube around it, which is represented as larger. How do things change when the observer of both arrangements (the three and the four-dimensional cube) moves?

When we go around the three-dimensional cube in a half circle, so to speak, and then look at the surface which was at first the farther as the nearer, this is now larger than the other across from it. By analogous conclusion, an inversion of the larger/smaller ratio of the inner and outer (three-dimensional) cubes in the four-dimensional "cube," is called an approach to the inner cube "in the fourth dimension." One would then have a connection of the positional relation to the two other initial interpretations (the fourth to length, breadth and height, or the fourth orthogonal, since in a four-dimensional cube four edges come together in every "corner point" which are, pairwise, perpendicular to each other).

19 Ibid, p. 55

By forming analogies between three and four-dimensional "cubes" and their connection with the arguments of orthogonality and position, Brotman comes to the conclusion that "two cubes of our real world, enclosed by one another," would be "described as equally large, but variously far away in the application of a four-dimensional language."[20] This, and the necessity of no longer labelling as such that which we normally call a "distortion" (in the case of the orthogonal rods), prove this four-dimensional language to be "too rich." What is undoubtedly meant by "too rich" is that this language offers additional descriptive possibilities to the ones already available, without being able to name the additional causes for this (e.g., perceived states of affairs). (We will not go into Brotman's explanation of the asymmetry between two- and four-dimensional language, because the prior considerations are already in need of critique.)

Brotman's essay closes with a remarkable focusing of her concern. It is granted that three- and four-dimensional "objects" (bodies are obviously meant here) were not talked about, but only imagined lines and surfaces, hence, geometrical forms: "I have only attempted to show how such a world could be *conceived*... "[21] In other words, proceeding from an extra-scientific pre-understanding of geometrical words for the description of a cube, Brotman discusses a fictive analogy for the four-dimensional case, leaps from there to the plane of talk about real objects through a "physicalistic interpretation" ("As far as I can see, the only possible physicalistic interpretation of surfaces, lines and points which comes into question is something like purely visual perceptions."[22]) and then verifies for the two and four-dimensional instances a collision with several conventions of our ordinary language. The conclusive reason for the rejection of two- and four-dimensional "conceptions," which can, of themselves, be consistently extracted, by analogy, from the three-dimensional, is ordinary language. Space is accordingly three-dimensional, because we always describe it so, indeed, the words "space" and "three-dimensional" have no other meaning except for precisely this, that our ordinary language distinguishes the number three for bodies, straight rods and positional relations. The question of whether this is done for good reasons, or *qua* error of our language-creating ancestors, is not considered.

Aside from this language-philosophical premise, a second characteristic of Brotman's proves her to be a typical representative of Analytical Philosophy. What the words point, line and surface mean in geometry, that is, how they are appropriately defined so that

20 Ibid, p. 60
21 Ibid, p. 62
22 Ibid.

one might, for instance, judge how they are to be applied to "objects," i.e., bodies, remains undiscussed - not coincidentally, but as a part of her own philosophy of science: since, in the end, these words are "interpreted" ("physicalistically") via "visual perceptions." What "looks straight" is "straight" and therefore means "straight" - only the physicist with theses about the straightness of light rays and the function of our eyes could explain this.

With that, a threefold critique of this proposal is suggested. (The possibility of an immanent critique is therefore not gone into):

1. No foundation for three-dimensionality is given, indeed, the author herself would probably not claim to have named reasons for three-dimensionality. The basis of the argumentation is the fact of our ordinary speech habits (and those taken into the ordinary language from science). A two-dimensional language is too poor relative to this, and a four-dimensional language too rich. The *fact* of the spoken language becomes the norm. The philosophical method limits itself to an analysis of this fact.

Whether this three-dimensional language can be *justified* (for reasons that must then be named), or whether it shows deficiencies, such as ambiguities, inconsistencies, redundancies or something similar, cannot be asked. "Normal" language is the final authority, and for this discussion a language is necessary which can be critically (normatively) revised when this is called for. Seen benevolently, Brotman's article is, at best, a piece of language analysis which shows what might be meant with the thesis that our ordinary language distinguishes the number three for spatial dimensions. This leads to a second point of criticism, which goes beyond the observation of abstinence from standardization (although standardization must, indeed, always be assumed):

2. The two- and four-dimensional languages appear in Brotman's analysis to be, in a certain sense, worse than the three-dimensional. When the two-dimensional language is labelled "too poor" (which is, of course, a pejorative expression), this is because the disappearance of a body (seen as a surface on the plane of projection) "behind" others - for various observers and at various times - cannot be expressed without talking about the passing away and regeneration of this body. It is precisely the latter - because of the difficulty of thinking up causal hypotheses for the causes of the disappearance and formation of bodies - which makes the two-dimensional language more complicated than the three-dimensional. Correspondingly, the four-dimensional fiction leads to a more complicated language. One describes states of affairs like the perceived size differences of cubes which are equally large, but at different distances, with the introduction of a new dimension. In other words, one invents two, more complicated alternatives to the existing language, and then is

happy for the simplicity of the original language relative to the others. In order to dignify this procedure, one must not let oneself be irritated by the difficulty of non-intuitive analogy formations. The layman may recognize this linguistic emperor's new clothes as nakedness, as an old tale and its common-sense understanding will show: A madman carries a heavy stone around for fear of lions. When asked what he would do with the stone if a lion showed up, he answers that he would just drop the stone so that he could run faster. The invention (and relieved surrender) of more complicated alternatives remains a cheap trick to make the initial situation appear simple, good, justified, or otherwise positive.

3. The third objection concerns the way in which Brotman uses geometrical words. The two- and four-dimensional fictions are allowed as "conceptions," because three-dimensionality is only an empirical fact. What we know from experience, must also be able to be (or conceived to be) principally different. Three-dimensionality can only be empirical when, in the case of exactly three rods which can stand orthogonally pairwise to each other, for example, all the necessary words like "straight" and "orthogonal" are already "operationally" defined. But as her reason for the *operationality* of her definitions, Brotman only indicates the fact that reference is made to physical "objects" (rods) and not to mathematical objects (lines). The entire remainder is left to the language competence of the speaker of ordinary language. "How a straight rod looks," or, that a "figure" is formed with two straight rods (lying orthogonally on each other) which "looks symmetrical" are her two examples. A simple objection must be made here.

We know how one can look at a wire in the form of an arc so that it looks straight, or at two straight rods which intersect at an oblique angle so that we see a symmetrical figure. If we take competence in ordinary language more seriously, like that of a carpenter, we immediately see that he does *not* only simply look when he wants to determine (or produce) straightness or orthogonality. Instead, he actually proceeds (more strictly than Brotman) "operationally," that is, he performs operations, like the construction of a perpendicular to the straight edge of a board, for which he needs only a slat (not even straight, but only sufficiently form-constant). And the carpenter is factually convinced that *all* right angles constructed in this way are equal. So that, for example, doors and windows which are constructed in this way will fit into a framework which was also constructed in this way. This technique is not only usual for craftsmen, but also represents the foundation of measuring physics. It is, in any case, a philosophical fairy tale that physicists define fundamental spatial forms as visual perceptions. Instead they order straight bars from workshops (for optical experiments, for instance) on which they pose "centered" light sources, lenses and screens, and then arrive at empirical theories about geometrical optics.

In other words, Brotman's interpretation of physics and her understanding of geometry have nothing to do with the actual competence of the speaker of ordinary language or with the actual procedure of physicists. Instead, Brotman is arrested here by a philosophy which is caught in the net of its own postulate. Namely, the postulate that every sentence is true when based either on definitions (analytically), or else based on experiences (empirically). Since it cannot be seen how three-dimensionality can be won purely analytically, it clearly must be empirical. Brotman's proof that three-dimensionality is won empirically fails because a more exact examination of the operational definitions of "straight" and "orthogonal" logically excludes the possibility of more than three pairwise orthogonal straight rods in one intersection.

THREE-DIMENSIONALITY AND CAUSALITY IN PHYSICS

Rudolf Carnap, the most important philosopher of the Vienna Circle, published a most original proposal[23] in 1924 in the *Annalen der Philosophie und philosophischen Kritik* which unjustly remains unmentioned in presentations of overviews of the three-dimensionality problem like those of Max Jammer[24] or G. J. Whitrow.[25] This proposal contains the thesis that the stipulation of the three-dimensionality of space is a logical consequence of the stipulation of certain causal relations in physics. Carnap's essay, which is also worth reading for other reasons, is a showpiece for proving how unsubstantiated the rejection of the Vienna Circle (at times, extremely polemic and malicious) by the established and, above all, historically-oriented, traditional philosophy in Germany was. His main thesis shall be introduced and critically discussed here. It will not be necessary, however, to represent Carnap's definitions (which are, in part, quite technically expansive) and the use of set theory. In his introduction, Carnap calls the assumption of the regularity of experience as well as the three-dimensionality of space, "fictions" and stresses that the object of his discussion is above all the thesis of their *connection*: "The fiction of the three-dimensionality of space (equivalent to the four-dimensionality of space-time) is the logical consequence of the regularity of all that happens."[26]

---

23 R. Carnap (1924)
24 M. Jammer (1960), pp. 193-208
25 G. J. Whitrow (1955/1956), pp. 25-31
26 R. Carnap (1924), p. 106

To provide a basis for this thesis, he first differentiates two levels of experience. The first is immediate sensory perception without the (as Kant asserted, "necessary") a priori forms. The second level of experience enters into formations through stipulation, e.g., the binding together of sensory impressions into "things" with "properties." Our everyday experiences of the world we live in, as well as the experiences of physics, are examples of experiences of the second level.

The structuring of the experience of the second level (Carnap speaks of the "*Umformung*" (transformation) of the first level of experience into that of the second level) in an everyday sense - in which things do something and are acted upon, and events are understood in terms of cause and effect - is much more complex and harder to understand than second-level structuring in physics, which occurs, as it were, within a uniform, simple scheme. Physics:

> knows no causality relations in the sense of an effect, and, in its purest form, also no substantiality. It constructs a world free of sensory qualities, in which there are only magnitudes of time and space, and certain non-sensory magnitudes of state. In the purest form, even these three types of magnitude do not have a character comparable to spatiality, temporality, or sensory qualities, but are merely numerical determinations, that is, relational terms.[27]

In other words, physics defines perimeters (magnitudes) through measuring procedures and formulates equations in which these perimeters occur in functional dependence to each another. The interpretation of such perimeters as spatial and temporal, or such functional dependencies as causal, merely has an illustrative character; indeed, the entire science of physics is, viewed correctly, "nothing but fiction."

The two levels of experience correspond to a primary and a secondary world. The primary world is that of the not yet interpreted sensory perceptions, and the secondary one is the interpreted world, the everyday or physically described world. Carnap's thesis can now be precisely formulated: In the transition from the primary to the secondary world, we must raise the dimension number (DN) by 1 in order to maintain the causal structure of the secondary world. (The reference here is always to 2+1 or 3+1 dimensions, by which the term "+1" means time, consistently one-dimensional in the primary and secondary worlds, and is not further discussed. Hence, only the raising of the DN of space is considered.)

Carnap undertakes an extensive definition and explanation of a set theoretical concept of the DN. To understand his argument, it

27 Ibid, p. 107

suffices to see that a non-spatial concept of dimension is established here in two steps:

1. "Class" is defined through a equality relation of elements (hence, tones which are judged by their pitch form their own class, just as tones which are judged by pitch and volume. In other words, the complete specification of the criteria which are to determine if two objects are equal, defines a class of these objects.)
2. For the elements of a class, a relation must be defined as "neighboring." With these two terms, a type of inductive definition of the DN of a realm can be given, first in one of its elements, then globally. This is presented here in the form of an illustration:

An example of a class is a mosaic, whose stones are the elements of the class. If the mosaic is made up of only one stone, and thus has no neighbors, then it has the DN = 0. The mosaic has the DN = n+1 in a particular stone *x*, when every stone which does not directly neighbor *x* is "divided" by a subclass DZ = n. Imagine a chain of mosaic stones comparable to a pearl necklace: every two stones which do not neighbor each other would be "divided" by at least one stone (which alone, for itself, has the DN = 0), that is, at least one (zero-dimensional) stone lies between them. If one considers a planar mosaic and therein a stone *x* and a non-neighboring stone *y*, these two can only be divided by a (one-dimensional) chain, i.e., the mosaic has the DN = 2. The class of mosaic stones which fill a crate take on the DN = 3, since two non-neighboring stones can only be divided by a subclass of the DN = 2, thus, a surface, etc. (Since the Carnapian definition is formal and general, and is not given with mosaic stones, it does not stop at 3, as is intuitively the case for mosaic stones.)

The example of the mosaic stones (which comes from Carnap) fits directly in the realm of visual perceptions: the visual field as a set of uninterpreted impressions is made up of color spots, like a mosaic, and is thus (spatially) two-dimensional. Following this, Carnap discusses the sense of touch, of hearing, and the muscular and static senses, and shows that even the interaction of several senses (like sight and touch) does not result in a DN greater than 2 for the primary world of uninterpreted sensory perception.

For the secondary, physical world, Carnap now defines various forms of laws; the so-called determining and limiting laws. The point of this differentiation for the following discussion is to show that physics (deterministic, not quantum theory) uses a space-time model in which three spatial dimensions are assumed homogeneous as possible propagation. Effects propagate - and this is a free stipulation of physics - not as *actio in distans*, "distant effect," but from space-time point to neighboring space-time point.

The argument which makes the raising of the DN in transition from the primary to the secondary world so cogent comes to this: we can perceive two spatially *divided* objects in our sensory, say visual, perception as *neighboring* color spots (given a suitable position to the observer, in any case). We are first forced to extend the two-dimensional world of sensory perception to a three-dimensional, constructed world, in that we only deem a causal interaction to be possible between objects which are neighbors (according to sensory perception) when the interval which lies between them is also included in the propagation. One can illustrate this as follows: Two houses, standing apart, can be seen by an observer in such a way that the retina image of the houses touch one another - when one house partially blocks the other, for instance. A causal effect between the two can only be assumed when the interval between the houses is also taken into account. The three-dimensionality of space is thus a consequence of our conception (or "fiction") of a homogeneous propagation, as we describe it with equations within the frame of the physical space-time model (Minkowski-World).

The remarkable thing about Carnap's proposal is that he does not attempt to derive the three-dimensionality of space, common as well as physicalistic, from principles or empirical propositions. Instead, he rejects such a derivation. "The number three of spatial dimensions ... is not known a posteriori, because it is not known at all, but decided, chosen. ... Nobody doubts that the secondary world, the common as well as the physicalistic, is (3+1)-dimensional."[28]

Hence, Carnap does not derive the three-dimensionality of space from any other givens, not even from those of descriptions of the primary world, but presents an argument that (with an explicit elucidation of the concept of the DN) *assumes* the following to be true:

1. The primary world has the DN 2;
2. the secondary (common as well as physicalistic) world has the DN 3;
3. physics assumes a homogeneous propagation (as the intuitive interpretation of the form of its mathematical theorems).

In reference to these three assumptions, the assertion which Carnap then argues is stated: assumption 3 compels the raising of the DN in transition from 1 to 2. The Carnapian argumentation has no immanent errors as far as we can see. It is, however, difficult to judge the question as to which new insights are gained in the sense that the premises 1 through 3 are themselves questionable.

Above all, two steps prove to be problematic in Carnap's considerations. Namely, the characterization of the experience of the first level (and the "primary world" connected to it), and a pragmatic

28 Ibid, p. 117

deficit in the characterization of the transition from the primary to the secondary world of physics.

Let us first address the experience of the first level and the primary world. They are, or correspond to, sensory perceptions without an interpretation of their contents. The primary world is the "given" or "originated," which exists independent of possible human perception. To be sure, the positivistic critique of Kant has shown that far fewer conditions for the possibility of perception are necessarily valid than Kant believed, but a total perceptual chaos does not reign before all intuitive interpretations, either. The color spots in visual perception being simultaneously perceived in the visual field would have, with necessity, spatial contact. *Proceeding* from this experience of the first level, or from this primary world, whose DN was shown to be 2, a *raising* of the DN by one must now follow, when a point-to-point causality is assumed, according to Carnap's main thesis.

But what kind of transition or "transformation" leads directly from uninterpreted sensory perceptions to the conceptual model of physical theories, i.e., to a four-dimensional space-time manifold as a homogeneous set of points? Does this transition correspond to something in real physics; not just a philosophical characterization of certain properties of theories, but the physics of the physicists? Presumably not. No one actually executes this "transformation." No one - capable of only sensory perceptions - could acquire or do physics without having first sufficiently mastered everyday life, i.e., no one who does not already have access to the common "secondary" world of Carnap's. The Carnapian "transformation" of the primary world into the secondary, *physical* world, which, in fact, makes the raising of the DN of space necessary, does not correspond to any real operation in the cognition process.

But could not Carnap's thesis be rescued by understanding this transition from sensory perception to physical theory, itself, as epistemological fiction (or reconstruction), in such a way that an analysis of the physics which works with three spatial dimensions would show that a sensory perception of two spatial dimensions forms its basis? Apart from the fact that it would be more appropriate to speak of *lowering* the DN by one in transition from the physical world to sensory perception, this rescue attempt does not help the thesis. There can be no reconstruction of the physics of the physicists which connects pure, uninterpreted sensory perception (as a binding link to the existing world, as it were) with the theoretical conception of a four-dimensional set of points as a homogeneous, physical space of point events. Because, as always, physicists still deal with such objects as falling stones, heated gases, charged metal plates, etc., and to this end they need instruments like measuring rods, clocks, scales, thermometers, etc. That means, even if it could be formulated with theoretical purity, which states that physics only deals with punctiform events in a (3+1)-dimensional manifold (quantum physics would no longer come into consideration then anyway), the physicists must nevertheless engage the things and properties of the world of day-to-day living in an epistemologically decisive "interaction" if they want to find out anything at all about the world. It is precisely the technical success of, for instance, the art of length and time measurement, the determination of the magnitude and position of bodies in three measures (length, height and breadth), and the creation of one-dimensional time on the face of a clock, which makes physical space (3+1)-dimensional, independent of any hypotheses or freely chosen fiction that only laws of close-range action may occur in physics. Or, stated in one simple sentence: without everyday experience, without the secondary world of the mundane type, in Carnap's sense, no connection between sensory perception and physical theory could be produced. The everyday world as the foundation of a high scientific stylization of experience and theory - fully acknowledged and undisputed in today's philosophy of science as being indispensable, by the way - also cannot be harmoniously introduced into Carnap's concept. True, Carnap says guardedly that the procedures of object constitution in everyday experience are extremely complicated, and that he will therefore have nothing to do with "strong" theses. Nevertheless, he assumes that the everyday world is also "secondary" to the primary world of uninterpreted sensory perceptions (analogous to the formation of the secondary world of physics). Which sequence can Carnap mean here?

A layman needs his own linguistic training to be able to communicate his uninterpreted sensory perceptions. To stay with visual perceptions, a layman can, upon inquiry, communicate what he sees. To do this, he will naturally use words for objects and their properties. Only with great effort will he be able to acknowledge that his visual field (in any given moment) is a closed surface made up of various color spots. In other words, the primary for the layman is not the uninterpreted sensory reception, but the perception. The first is a theoretical construct which is formed through the epistemological discussion about perceptions, their contents and their origin through the function of our sensory organs. The primary given for the person who does not yet philosophize or theorize is, contrary to Carnap, the perceived, which one must master from the beginning in order to be able to know and say something about sensory perceptions (and to determine anything about them).

Since Carnap stands the sequence of epistemological steps in sensory perception/sensory reception on its head, the raising of the dimension number of space in the formation of perception from sensory receptions is also a questionable concept. As humans, we learn perceptions in connection with language acquisition. With our actions, we participate in a social context and, in the process, we already learn a three-dimensional spatial language of perception. "Nobody doubts" that the space of everyday life is three-dimensional. *Proceeding from this* it is then possible with our perceptional concept of dimension, for which an oil painting or a mosaic is two-dimensional, to set up a theory of sensory perception, in which a *lowering* of the DN to one occurs. And *relative to this theory* - today within the framework of artificial intelligence research, for instance - it can be asked how our organism manages (by means of theories about machines, already in terms of theoretical physics) to provide orientation in a three-dimensional world with sense organs that work with two-dimensions. A similar question, for instance, is how a mobile robot, equipped with a camera, can take targeted objects in its claw and lay them on a prescribed place.

In summary, Carnap's attempt (although well worth reading) must be appraised as unsuccessful. The main thesis of a logical connection between the two fictions of the three-dimensionality of space and causality principles is not proven, because his reconstruction as a transformation of sensory perceptions into a knowledge of natural laws fails. However, Carnap did clearly see that three-dimensionality is neither necessary a priori, nor empirically, but that it is based on stipulations, which, for their part, are also in need of a foundation. Properties which are picked from physical theories in the latest levels of development cannot provide a foundational argument like Carnap attempts here. Instead, only those properties which make a methodological physics possible in the first place (whose premises with respect to space must be acknowledged so that a measuring, experimental physics is possible) or else of spatial orientation, at the level of everyday experience (finding one's way through the woods, or a city; characterizing bodies in regard to form, position and magnitude, etc.).

On the whole, however, attempts (which characteristically originate from representatives of Analytical Philosophy) to retrospectively present the three-dimensionality of space as a stipulation which can be presented as a necessary presupposition *after* a successful *causal* description of the world, have not been successful.

CHAPTER FIVE

The Biological and Perception-Theoretical Approaches

PRELIMINARY REMARKS

All four of the previously discussed groups of approaches, the purely spatial, the kinematic, the dynamic and the causalistic, attempt to provide a basis for the three-dimensionality of space - at least to the extent that they do not belong to those rather ridiculous groups of approaches which play around with numerical speculations - through reference to properties of the outer world, of bodies and their motions, to forces in effect between bodies, or the causality which is found in the world of bodies. The fifth and last group, which shall now be discussed, does not search for reasons in the *objects* of human knowledge, but in the *subject* which gains knowledge itself. In this, such things are considered as the construction of our sensory organs, the biologically determined structure of our sensory perception, the construction of our organs of equilibrium, the interaction between perception and the motion of the perceiving subject and, finally, the thesis of evolutionary adaption of organisms to their environment, which can be recognized as a three-dimensional environment with the capabilities of organs selected through natural history.

This fifth group cannot be completely separated from those prior to it. The reflective representatives of the previously discussed approaches always at least mention perceptional conditions, and some even explicitly take them as a theme. Especially the analytical approaches of the fourth group - which even emphasize the linguistic form of human knowledge and stipulations guided by aspects of purpose and function - cannot be sharply divided from the perception-theoretical attempts. The closest contact is, of course, between Carnap's proposal and the theory of Poincaré which will be discussed within the fifth group. They both give a general (i.e., not spatially specified) definition of "dimension" in order to then apply this to the physical world, and sensory perception plays a decisive role in this. However, the decision to treat Poincaré's approach as a perception-theoretical one could, perhaps, be justified by a difference in principle between him and Carnap. Poincaré does not primarily make reference to physics, but treats "space" as concept formation in connection with perception, or, more precisely, as the formation of a physical continuum. The possibility that physical theories could distinguish the triad of spatial dimensions only plays a role for Poincaré to the extent that, in physics, the motion of the limbs of the human body can be described just as well as rigid bodies. In any case, the result is that Poincaré fixes three spatial axes for translation motions and three for rotations.

Poincaré is also one of the most important authors on the subject of the three-dimensionality of space for other reasons. He is not only the founder of topology (the mathematical discipline which is accountable today, *ex officio*, for the definition of the concept of dimension) but he expressly developed this discipline with the goal of demonstrating the *three-dimensionality of empirical space* (in two different ways).[29]

The dimension number of perceptual space was a subject which had only been incidentally treated, when the rise of physiology and experimental psychology, especially in the nineteenth century, brought about a broad discussion of the structure of perceptual space. The actual issue at question was whether perceptual space is Euclidean or non-Euclidean. Even today, this question comes up again and again, especially since the turbulent development of perceptual psychology in the last thirty years. From an epistemological perspective, however, these newer approaches are less productive because they make use of physics and its (3+1)-dimensional description of nature without any awareness of the problematic. Today psychology and physiology are to the greatest extent, physicalistic and naturalistic. That means that theories of perception assume the availability of knowledge about what there is to be perceived. When in doubt, physics decides with disciplines like optics, which provides an objective description of the environmental signals for visual perception - hence, "physicalistic." When not even these scruples arise, the perception-theorician considers it to be nature itself, existing independent of humans, which is perceived by humans - hence, "naturalistic." Humans perceive nature, in principle, exactly as it is perceived by animals, although in a more complex way, perhaps. Regarding these perception-theoretical approaches, it must now only be clarified in a general way, whether they can accomplish more than the transposition of the assumed dimension number of a given space to the construction and function of the sensory organs.

A critical revision of these passivistic approaches (passivistic because the perceiving organism is misunderstood as a passive receiver of signals) was presented by Jacob von Uexküll in the concept of the *Funktionskreis* (cycle of function), which assumes a interaction between the *Merkwelt* (perceptual world) and the *Wirkwelt* (world acted upon). This means the limitation of the characteristics of the outer world to the reaction and action possibilities of an organism. The question here is whether, according to this premise,

---

29 H. Poincaré (1912), pp. 483-504

the three-dimensionality of space is limited to *human* perception (and thus "more primitive" animals perceive and act in spaces of lower dimensions), and whether this theory can be free of anthropomorphic premises.

In conclusion, because of the attention given to the alleged replacement of the philosophical theory of perception by a "biology of perception" or "evolutionary theory of perception," we will speak of the explicitly stated theses found therein about the three-dimensionality of empirical space as the field of natural-historical trial in the struggle for the survival of the species.

THE PHYSICAL CONTINUUM VERSUS THE MATHEMATICAL CONTINUUM (HENRI POINCARÉ)

Henri Poincaré, equally important as a mathematician and as a philosopher of science, repeatedly occupied himself with the problem of the three-dimensionality of space, as in a particular chapter of his book *The Value of Science* (French publication in 1905, English translation in 1913), or in the (posthumously published) essay, "*Pourquoi l'espace a trois dimensions*." We will not go into marginal differences between the various texts (and a possible development of his perspective) here. Consistent throughout his work is his distinction between a mathematical space, and a physical one, which is constituted through our sensory perception. In this, the mathematical consideration provides the definitions which are then applied to the physical continuum.

Poincaré once again picks up the expression (already used by Leibniz) "*analysis situs*" (analysis of the position, namely of geometrical forms like point, line and surface) in order to characterize the part of geometry which forms the "real geometrical intuition." This is the basis of the higher branches of geometry. If geometry "is the art of good considerations of badly made figures," then a artist must not neglect certain properties, otherwise his drawings become unusable, and the geometrical intuition injured. For example, he must "not render a closed curve as an open one, or three lines which intersect in a point as three which have no point in common, or a surface which has an opening as a surface without an opening."[30] While metrical geometry is about magnitudes (equality or ratios of segments and angles), and projective geometry is about projective invariants (properties of figures which remain the same regardless of the perspective of the observer), *analysis situs* is concerned with more fundamental properties, independent of the

30 Ibid, p. 484

magnitude or form, like the closedness or openness of lines or surfaces. In metrical geometry, two similar or two congruent triangles are equal in a sense which is described by magnitude specifications. And in projective geometry, a rectangle is equivalent to the quadrangle which is seen its shadow on a wall which is not parallel to it (straight lines remain straight lines; corner points remain corner points). But in *analysis situs* a triangle and a circle, or a cube surface and a tetrahedron surface, are equivalent. The criterion, in reference to which equivalency can be determined, is the closedness or openness of lines or surfaces, or more generally, the requirement that one can be produced from another "through a continuous shape-change." The decisive aspect here lies in the "shapelessness" of space, which has no metrical or projective properties. This is just another way for Poincaré to say that in this space, figures are classified in such a way that neither specifications of magnitude, nor of form, play a role. The mathematical layman can illustrate these "topological" properties for himself, if he thinks of a figure which is printed on a balloon; however one stretches or twists it - without ripping it, of course - two lines which intersect, for instance, will always meet in an intersection, and one will not be able to make an 8, i.e., a closed figure with a point of intersection, from a simple closed figure like a circle.

Mathematicians discuss such deformations of figures arithmetically, that is, they represent points with one or more numbers, and their relative dislocation in the deformation of a figure with functions (familiar to the layman from analytical geometry lessons in school). Functions which, for example, "deform" a segment into a surface, or vice versa, a surface into a segment, are clearly possible, but are not "continuous." Continuous transformation, in contrast, is a condition for the topological equivalence of figures. A sphere can continuously be transformed into a cube in the sense of the surface as well as in the sense of the space-filling body, but a sphere cannot be continuously transformed into a ring or a circular surface area. This condition of continuity, whose mathematical meaning need not be gone into here, in order to exhibit its application for the physical continuum, is the reason why Poincaré speaks of "continuum" here. A mathematical continuum is a continuous set of points.

Then "cuts" are defined for figures in topological space. For a start, the literal meaning of "cut" - like that of a knife cutting through an apple - is not misleading. Really splitting something into parts is what is meant by cut, as opposed to a cut through a ring which fully cuts through the ring, but which does not lead to its halving. We need not go into the necessary case distinctions of topology (the word commonly used today, in place of "*analysis situs*"); the layman could simply consider in how many ways, and with which results, he could cut up a pretzel. Poincaré's considerations are only in regard to simple forms, like a segment, a (e.g., circular) surface without holes and a body, or spatial figure like a sphere, likewise without holes. Two different points of these figures, *P1* and *P2*, could then be completely separated by a cut (that is, so that there is no connecting line within the figure for *P1* and *P2*), when, lying between *P1* and *P2*, there is, 1) in the case of a linear area, another point, 2) in the case of a circular area, a line from boundary to boundary, and 3) in the case of a sphere, a surface whose boundaries' edges lie completely on the sphere's surface. If, however, we want to separate two different points inside a sphere with a line, we would be unsuccessful, because a connecting line between the two points can be found which does not meet the dividing line, and which lies completely within the sphere.

What is meant by "cut" here - now only in a metaphorical sense - is any real division of a geometrical form into parts, such that every point of this form belongs to exactly one part.

Evidently the three examples of the division of the segment, the surface, and the body exhibit the well-known connection between the dimensions of the divider and the divided: the body is divided by a surface, the surface by a line, and the line by a point. The point itself is, in the mathematical continuum, "an individual, perfectly distinct from the others and, what's more, perfectly indivisible." Consequently, there can be no division of a point. The "recursive definition," according to which a division through an n-dimensional form has the dimension n-1 (with the stipulation of the dimension number zero for the point) arrives at the usual definitions of the line as one dimensional, the surface as two-dimensional, and space as three-dimensional.

Poincaré's procedure for the mathematical continuum differs from that of the ancient geometricians which we spoke of earlier (and Aristotle's corresponding argument). They always thought of *homogeneous* forms, hence, *flat* surfaces and *straight* or *circular* lines, where the connection of the divider and divided (in ancient terms, bounding and bounded) was considered. The change which the shift of the dimension problem experiences in topology through Poincaré is of far greater consequence than it might seem at first glance. Into the present day, it is self-evident that "dimension" is not a geometrical term, but a *topological* one which, it is true, is then used in analytical geometry, but which is thoroughly defined in topological spaces. What is epistemologically important is that the topological consideration is "shapeless" (Poincaré), i.e., that no form characteristics of geometrical figures can play a role in the definition of the concept of dimension. (In contrast, the thesis of this book is that a suitable definition of spatial dimensions can only succeed in form theory.)

This leap makes the great distance to intuitive talk of "cuts" clear. One can still argue using ordinary language that a cut through the corner of a cube leads to other corners, and that therefore - by linguistic stipulation - no new form is brought into play terminologically. In the topological consideration, however, talk of "corners" or "edges" is meaningless. The corner of a cube is topologically equivalent to a point on a sphere. *With that, the basis of plausibility for the indivisibility of the point, as well as its zero-dimensionality, falls away.*

The criticism against the recursive definition of "dimension" for mathematical continua, that the (two-dimensional) surface of the double cone is only divided by a single point, and is thus zero-dimensionally divided, was likewise made by an important mathematician (and philosophically important logician), Luitzen E. J. Brouwer. Brouwer himself gave another set theoretical definition of dimension which worked with congruence, and proved that the dimension number is an invariant for Euclidian space, i.e., that no n-dimensional space can be continuously transformed into a p-dimensional one ($p \neq n$).

We will leave this mathematical development because, as stressed at the beginning, it is precisely a general definition of dimension, which is not restricted to the number three, which is sought and found here. Even Poincaré does not claim to be able to prove three-dimensionality in connection with the mathematical continuum. Instead, "the *presupposition* of *analysis situs* is that space is a three-dimensional continuum."[31] The three-dimensionality of space is more an "empirical fact" in the realm of sensory perception (an assertion which directly contradicts Carnap's thesis).

The most important difference to the mathematical continuum is exhibited by the physical continuum concerning the equality and inequality of its elements: it is the barely perceptible difference (as between two bodies whose weight is estimated) which allows for the definition of a chain of homogeneous perceptions. The structure of such a chain is principally distinguished from a mathematical continuum of points to the extent that we are, for instance, capable of distinguishing between a 10 gram and a 12 gram weight when weighing them by hand, although neither is distinguishable from an 11 gram weight, while two different points could never be identical with a third. A physical continuum is then defined such that the following applies for such a chain, i.e., "a system of related elements ... [one can] consider any two given elements of a system as the end

31 Ibid, p. 485

links of the type of chain just discussed..., whose links as a whole belong to the system."[32] For example, if any two given weight estimates are connected by a chain of barely perceptible weight distinctions, then the weight estimate forms a physical continuum.

The concept of "cut" is now extended to this type of continua, and the number of its dimensions is determined in this way.

> Assuming that one excludes any given element of system S, and all those which one cannot distinguish from them. ... The totality of the isolated elements form a cut. ... Now a continuum which one can split into parts, in that one isolates a finite number of elements, will be one-dimensional. A physical continuum will have n dimensions when one can split it into separate parts in which one leads sections which are themselves continua with n-1 dimensions.[33]

Hence, after clarifying what is to be understood as a physical continuum by referring to the Fechner law, or the concept of the barely perceptible distinction, Poincaré then turns to the senses of sight and touch. In this he leans upon empirical facts of the physiology of sense reception: in visual perception a two-dimensional picture originates in the retina which fulfills the condition of being a physical continuum. Then depth vision is explained by the convergence of the eye axes, but also through the accommodation of the eye lenses. If both depth estimates (experienced as internal muscular sensation) did not coincide, we would have to speak of a four-dimensional visual continuum. Since they do coincide, however, a three-dimensional continuum can be assumed. For the sense of touch, Poincaré discusses the definition of elements as barely distinguishable local stimuli (which, measured in centimeters, are different for different areas of skin) and arrives at a two-dimensional continuum ("just like the surface of the skin").

In both cases, the sense of sight as well as the sense of touch, several "abstractions" are drawn, i.e., one does not take certain things into account, like the brightness of the optical stimulus, or the pressure or kind (dull or pointed) of touch stimulus (only simultaneous receptions are considered in any case). If such stimulus qualities were taken into account, the criterion of the equality or inequality of elements of a physical continuum would be richer, and its dimension number accordingly higher.

In any case, the question of whether "space" has already been achieved through this characterization of the senses of sight and touch, meets with a decisive "no" from Poincaré. Thus, it is *not* the description of the sensory organs (two-dimensional retina or skin, and

---

32 Ibid, p. 489
33 Ibid, pp. 489-490

a depth dimension through axes convergence) which allows a statement about the dimension number of space. According to Poincaré, it is rather the case that a relationship between the sensory organs and the external world must first be produced, which is solely possible through motion.

To this end, "we must first distinguish between external changes, which do not depend on our will and are not accompanied by muscular sensations, and internal changes, which are motions of our body and which we can distinguish from the former in that they are voluntary and are accompanied by muscular sensations."[34] The external motions (like that of a moved body) are divided between changes of "position" and changes of "state," according to the criterion that only *displacements* of things in the external world can be "corrected" by internal motions, that is, motions of one's own body. One follows the motion of a thing with the eyes, in order to bring the picture of this thing "to the same point on the retina." It is not entirely clear if Poincaré means rotational motions (which one can at least correct by going around the rotating object), or the displacement of parts of bodies (as when an animal moves its head), when he speaks of changes of attitude which cannot be corrected by the perceiver's internal motion. This lack of clarity does not, however, play a very important role since, according to Poincaré, it is finally only muscular sensation, which a person has as a correction for the displacement of things in the external world, that produces the long-sought connection between the internal and external world - at least regarding the sense of sight.

Things are different regarding the sense of touch. When a finger touches a stone, for instance, a two-dimensional continuum results because of the two-dimensional surface of the stone (and not the surface of the human skin!). As a provisional result then, visual and tactile spaces do not coincide.

"Space," as the space of our sensory experience to which we apply geometrical intuition, does not, however, come about through isolated events of sight and touch; after all, people born blind also develop a three-dimensional concept of space. Instead, space is first built up from the interaction of the senses of sight and touch in motion - exclusively on the basis of personal muscular sensation. Visual and tactile space coincide in a learned, practiced human ability. One touches a body which one simultaneously sees. Now this body withdraws so that the touch contact with the finger is broken. We then learn to produce this tactile sensation again by moving the finger along the visually perceived depth. Hence, tactile space gains

---

34 Ibid, pp. 493-494

its third dimension through compensation for the loss of touch experience as it corresponds to visual depth-perception.

With that Poincaré "arrived at the definition of a continuum which is a real space; it is the space which is thought to be described by one of my fingers."[35] (We will pass over Poincaré's demonstration that the spaces which are described by the other fingers are isomorphic with that of the first, and therefore have the same number of dimensions.) Hence, the goal of Poincaré's considerations has been achieved. When the dimension number "of space," as the space of our daily (sensory) experience, is considered, then neither measurements nor magnitudes (as in metrical geometry) nor shapes or forms (as in projective geometry) matter terminologically. Only its shapeless continuity structure matters. The "recursive" definition of dimension via sections could be carried over to the physical continua, if appropriate criteria for the identification of elements of a chain of homogeneous sensory impressions were given. The physical continua of the visual and tactile senses interact through changes which are compensatory motions for the displacement of things in the external world, and finally form three-dimensional space solely through the (internal) muscular sensations.

What knowledge - aside from the conceptual apparatus of topology - does Poincaré invest? And which silent stipulations are indispensable for his theory? We will begin with the latter question, following the construction of this theory:

Talk of continua and cuts, and therefore also a definition of dimension, is certainly possible within the framework of a formal topology without any reference to spatial intuition, that is, without terminological explanations which make reference to drawing practices (lines on surfaces; points as intersections of two lines). (We need not discuss Brouwer's immanent criticism again here.) But in dispensing with the known (ordinary language) meaning of such words as surface, line and point, it is not possible to see how we can know that the retina and the skin of a body are surfaces, or that a thing in the external world, like a stone, has a "surface." The leap from an abstract space and dimension theory to its application must be explicitly justified linguistically. Of course, Poincaré proceeds *de facto* as follows: he, as a mathematician, knows the role of invariants for the characterization of individual branches of geometry, so he then begins with Euclidean geometry (which is a geometry of similarity, and thus formulates invariants in terms of magnitude) and goes through projective geometry in order to achieve a more general mathematical "geometry" - one in which only the continuous transformation (shape-change) of figures defines figure equality. This

35 Ibid, p. 495

can be illustrated by the practice of drawing (a meaningful type of action which is practiced with specifiable purposes and which can be described in statements which are capable of consensus). And this technically successful practice still provides the means for presenting the physical continua, too.

All this has not yet formulated a criticism against Poincaré. The premises have only been named which will be important for the following judgment of Poincaré's theory.

Poincaré repeatedly stresses that "experimental facts" give occasion for the assumption of three-dimensionality. These are, on one hand, the facts of sensory physiology and perceptual psychology (where details of visual or tactile perception are spoken of), and on the other hand, facts of physics. Because one must know something about the possible bodily motions (which are carried out during perception in compensation for motions of the external world). To this end, Poincaré takes the following *from physics*: "Now experience shows us that the choice of these coordinates [in which physical laws are formulated] has no influence, and that the equations are not changed by a change of axes. The entirety of all possible axial changes forms, as it is expressed, a group of six dimensions."[36] Or more briefly, in *The Value of Science*:[37] "Now how do we know that this continuum of displacements has six dimensions? We know it *by experience*." (Poincaré's italics). The six motions, namely three translations and three rotations, correspond to the triad of spatial axes, and thus to the three dimensions.

> Now since this group is also isomorph with the group of motions of our limbs when one considers them as solid bodies, and since the properties of solid bodies, which obey the laws of this group in their motions, are, in the end, only a special case of the general invariance properties which we have alluded to, one sees that there is no *essential* [Poincaré's italics] distinction between the physical reasons which compel us to attribute three dimensions to space, and the psychological reasons which ... are set forth [here].[38]

This still does not make Poincaré an empiricist or a physicalist, since he stresses that the triad of spatial dimensions in physics is a matter of conventional choice and theoretical comfort. But he invests several assumptions in this which are neither trivial nor cautious, but which only seem acceptable from the foundation of accepted physical theories:

---

36 Ibid, p. 501
37 H. Poincaré (1958), p. 57
38 H. Poincaré (1912), p. 502

1. Poincaré holds it to be an empirical fact that physical laws are invariant, particularly in reference to three-dimensional spatial-positional descriptions of events. There can be absolutely no doubt, however, that nobody knows or has attempted an experimental practice which operates with two, four or more dimensions. The invested geometry and the invested measuring techniques are dependent on Euclidean concepts (point, line, surface) as well. If it could then be shown, which Poincaré in no way accomplishes, that particular physical experiences lead to the acceptance of three-dimensionality, then it could be shown - in all examples of experiences which actually occur in physics - that three-dimensionality has already been introduced into them by virtue of definition and technical practice.

2. Empirical facts of the psychological or physiological type are also not accessible independent of a conceptual and technical structuring of space in the sense of a three-dimensional Euclidean geometry. What is meant by the "structuring of space" is that psychologists and physiologists of course use our normal geometrical intuition as a basis when they perform perception experiments or study the anatomy of the eye. Topology does not come to bear in this, but the assumption of Cartesian coordinates does - the *ad hoc* most convenient position of three planes, pairwise orthogonal to one another. How else could it be understood that we have different muscular sensations in depth-perception for near and far objects, if we did not know that for a constant eye distance the convergence of the eye axes must be bigger for the nearer object? And how could we speak of the muscular sensation of accommodation, if we did not know from physical optics that in focusing on a distant object the lens must be flatter, hence, extended?

3. Finally, Poincaré does not distinguish between the sensory perception that a person quite simply has when he sees and touches (and, in so doing, is unaware of the fact that he is perceiving), and a perception that has been sharpened by an attentiveness gained from perception theory (or even from the "experiences" that only occur in the theory of perception). It is a retrospective theoretical interpretation that when we observe an object which changes position, we follow it with our eyes and correct in this way for a motion in the external world, in order to bring the retinal image of the object to the same point on the retina once again. We describe our perceptual behavior in the same way as Poincaré, but we do not perceive the state of affairs that the retinal image of the object is in the same point on the retina. Nor can we have the experience that depth-perception results from axes convergence and accommodation. One can, indeed, produce an "unpleasant feeling" (like wearing glasses which are not prescribed, but through which one can

nevertheless see sharply) when one has accommodated to a square lattice, but the axes convergence does not fit, so that the eyes are adjusted in such a way that the two pictures are shifted against each other by just one interval.

In summary, Poincaré's acumen does not simply equate the dimension number of (external) space with the dimension number of our sensory organs, nor derive the former from the latter, but seeks a constitution of space through the interaction of our senses in (compensatory) motion. But this is precisely the main reason for the fact that, in the end, all that results from a theory here, is that which was already invested into it from the beginning as theoretical knowledge and technical practice. Precisely because scientific-experimental knowledge from physiology and psychology is needed, and precisely because the motions of the perceiving subject (whose muscular sensations are, according to Poincaré, the sole seat of conscious perception) are described with the conceptual means of mechanics, hence physics, a spatial conception which is at least projective or form-theoretical, perhaps even metric, is already explicitly used. Poincaré overlooked this, and that ultimately makes his proposals insufficient.

THE DIMENSION NUMBER OF THE *MERKWELT* AND THE *WIRKWELT* (JAKOB VON UEXKÜLL)

> We indulge ourselves all too easily in the delusion that the relationships of the foreign subject to the things of its environment take place in the same space and in the same time as the relationships which connect us to the things of our human world. This delusion is fed by the belief in the existence of a single world which all living beings are plugged into. The generally cherished conviction that there must be only one space and one time for all living creatures arises from this.[39]

In this and many similar passages, the biologist Jakob von Uexküll attacks the naivety of a silent assumption - widespread among layman as well as scientists - which is made when speaking about the perception and behavior of animals. To formulate this criticism differently: it becomes an unsolvable riddle as to how bees, for instance, or migratory birds learn and apply the astonishing geometrical and navigational theories which human scientists need to describe the accomplishments of these animals. According to Uexküll, the solution to this riddle lies in the fact that animals do not

39 J. v. Uexküll (1970), p. 16

move in the same space which originated with humans, through the cultural history of measurement, of geometry, of navigation in sea and air travel and, not least, through a sensory perception which has been transformed (by language) throughout cultural history.

Stated more simply, every animal only has the "space," and the corresponding perception of spatial states, which it can react to. More generally, i.e., not limited to spatial states, Uexküll asserts that the environment for animals has only those features or attributes for which the animal also has "*Wirkmale*" (activities), i.e., those to which it can react. "*Merkmale*" (features/attributes) and "*Wirkmale*" (activities) mutually condition themselves, and interact in a "*Funktionskreis*" (cycle of function). The set of *Merkmale* for an individual forms its "*Merkwelt*" (perceptual world), and the set of its (organic) possibilities for acting in and affecting the world is its "*Wirkwelt*" (world acted upon).

In his *Bilderbuch unsichtbarer Welten*[40] (Picture Book of Invisible Worlds) Uexküll illustrates this as follows:

> Just as a gastronomer picks only the raisins out of a cake, the tick extracts only the butyric acid from the things in its surroundings. We are not interested in knowing which taste receptions the raisins cause the gastronomer, but just the fact that the raisins become features of his environment because they are of particular biological significance to him; we also do not ask how the butyric acid tastes or smells to the tick, but we merely register the fact that, because biologically significant, butyric acid becomes a feature to the tick.[41]

This connection between *Merkwelt* and *Wirkwelt* is not only characteristic of animals, but of humans as well. Even in reference to space and the number of its dimensions, sensory perception as a passive, receptive capability is not the sole foundation for the formation of the concept of space.

Under the title "*Umwelträume*" (Environmental Spaces) Uexküll examines (in the following order): "*Wirkraum*" (space acted upon), tactile space, and visual space. To this end, he always makes comparisons between animals and humans. After a short presentation of his main arguments, we will discuss this theory immanently, as it were, in order to, as a second step, take a look at the matter "from the outside." While one must (with Uexküll) confess that nobody sees, or possesses "space" from the outside, but is ultimately always in his own "soap bubble space" (see below), one can nevertheless consider *theories* "from the outside," i.e., from another cognitive perspective.

40 The subtitle of the book by J. v. Uexküll (1970)
41 J. v. Uexküll (1970), p. 15

The *Wirkraum* is the range of the motion of our limbs. Not only the reach of our arms is meant here, but also the reach of our legs, that is, walking. All motions (in the terminology of Poincaré: all changes of position and attitude) are divided into three pairs of directions: to the right and to the left, above and below, and to the front and to the back. We refer all directional specifications to a "system of planes positioned perpendicularly to one another, the commonly known coordinate system which serves as the foundation for all spatial determinations" which lies firmly in our bodies. It is seen as an empirical fact by Uexküll that this is the case, of which one can easily convince oneself ("nothing easier than this") with an experiment: with eyes closed, one moves one's open hand horizontally up and down in front of one's face (and the eyes, or with some people the upper lip, will be recognized as the boundary between above and below), then vertically back and forth in front of one's face (and the middle of the face will be recognized as the boundary between left and right), and finally vertically, backwards and forwards at the side of the head (and the ear, or with some people, the cheekbone or the tip of the nose, will be recognized as the boundary between front and back).

Uexküll now relates this "coordinate system" (a misleading label here, because only three body planes - not coordinates, which are intervals of the points of intersection of these planes - are determined) to the construction of our organs of equilibrium, i.e., the three colonnades of the inner ear, "whose position approximately corresponds to the three planes of the *Wirkraum*." Then he proceeds, "This connection has been so clearly proven by numerous experiments, that we can make the assertion: All animals that possess three colonnades, also have a three-dimensional *Wirkraum* at their disposal." The fish is then described as an example of a being whose colonnades play the role of "a compass" according to Uexküll, but not one "that always points North, but a compass 'for the front door'. When all motions of the whole body are split up and marked in the colonnades, then the animal must find itself at its starting point when, in its wanderings, it has led the nervous marking back to zero."[42] So much for the *Wirkraum*.

In connection with tactile space, aside from a brief mention of the Weberian tactile experiments (two tactile stimuli, kept a constant distance apart, are perceived as two separate stimuli on some places of the skin, but as only one on others), only the special role of the tactile sense for the spatial orientation of certain animals (bees, mice and rats) is mentioned.

Finally, for the visual space, the phenomena of the "farthest plane" takes the foreground. This refers to that hemispherical surface

42 All quotations are taken from J. v. Uexküll (1970), pp. 16-19

beyond which objects can no longer be judged to be farther or nearer, but only bigger and smaller. What is discussed here is the increase of this farthest plane (it would be better to call it a sphere) which encloses every organism like a soap bubble, as it were, from the stage of a small child (approximately three meters) to the adult stage (six to eight kilometers). The problem of determining the farthest plane in animals through experiment and observation is also discussed. The dimension number is, at best, indirectly referred to in the sentence: "Like in tactile space, the connections from place to place are accomplished through directional steps." And for tactile space it is merely stated: "Places are connected by palpating with the directional steps, and both serve to render the form." In this context "places" refer to places on the epidermis, on which certain tactile receptions are localized, and "form" means the form of things in the environment, like that of a touched stone.

The three-dimensionality of space is, within the frame of the Uexküllian distinction, primarily that of *Wirkraum*, whereby, without further explanation, the triad of the perpendicularly paired planes through the human head is taken as the dimension number. (The fact that "*dimensio*" means diameter, and has something to do with measurements or measuring, is not felt to be any more disturbing here than in any other space or dimension theory. At least it can be said in favor of this terminological generosity, that one can also say in German that one "measures" (*durchmißt*) space or a segment when one travels down a path[43] - quite in the sense of the "*Wirkraum*.")

There are, however, two points in which the Uexküllian proposal is unclear, indeed, even vulnerable to attack. The first concerns the mentioned experiment with which every person should be able to determine a coordinate system within himself, and the other is the function of the three colonnades.

The experiment to exhibit our certain intuition of spatial directions, the same for every person, referring to one's own body, can *only* be carried out with people who understand the linguistic instructions which are given. In other words, whoever does not already know how to hold and move his hand, and whoever does not understand the questions about the boundaries of above/below, front/back and left/right, also cannot determine anything at all on himself. Only a person who is already bound to the language and to

43 Translator's note: The German word for diameter is "*Durchmesser*" and the related verb "*durchmessen*" literally means "to measure through," although this verb is more properly translated with "to stride across" in reference to space, and with "to cover" in reference to a segment.

the linguistically mediated spatial orientation can find "his" coordinate system.

But then it becomes a problem to say just what it is that this experiment is supposed to inform us about. Is it about a natural fact, or about how one learns the three word pairs for directions in English? Uexküll's argument ignores language here (as in all the other passages of his text as well), i.e., it does not even attempt to justify the meanings assigned to words, or their introductory or definitional contexts. Hence, the conventional portion of his experiment eludes him.

Let us consider, as evidence for this criticism, how these three word pairs are learned and used. The Uexküllian experiment assumes that this usage (and therefore the learning) only occurs in an "indicative" sense and not in a "predicative" sense. Indicative words, or indicators, are words which are accompanied, or linguistically replaced, by demonstrative gestures. "This," "here," "now," but also "front" (as in "always look in front of you when driving") and "left" (as in the comment of the tour guide "to the left you see St. Elizabeth's Cathedral") are demonstrative, indicative words. Predicative words, or predicators, are determined from examples and counter-examples. This is not only the case with color-words like "red" or "green," nouns like "house" and "oak," or verbs like "to go" or "to move," but also with words of position and direction. When a letter lies "on top" of a stack of letters, or when Heinrich Heine's blue ribbon lies in the back on the left-hand side ("*Sie gab bei ihrem Tod/ mir ein blaues Band./ Es liegt in der Kommode/ ganz hinten linker Hand*"), these are predicative designations. They can be determined from spatial states of affairs, which one can produce one's self as suitable examples for language acquisition. No reference is made in them to the inner coordinate system.

It is a question for developmental psychologists whether we first learn directional distinctions indicatively or predicatively in the acquisition of our native language. This is an empirical question and should not be decided by speculation here. But it is certain that we, as adults competent in our language, master such directional specifications indicatively as well as predicatively, without normally being aware of this difference. But then it is principally undistinguishable as to what it is that we answer to in Uexküll's experiment when we are requested to localize the said boundaries, because the indicative "front" is equivalent in meaning to the (predicative) "front" in reference to our own body. Thus, whenever linguistic problems of understanding come up, "front" must be terminologically reconstructed as a predicator, that is, using exhibitory or producible examples. The dependence on language, and therefore the dependence on an individual history of language acquisition, as well as linguistic rules and a community of speakers, is obvious. However, absolutely no spatial structure is derived or known through this intuitive stipulation, or determination, which is not already present in the system of linguistic distinctions of the experimenter and the language proficient subject of the experiment. In other words, the three-dimensionality of space is already contained in the triplet of word pairs for directions. If we imagine a natural tribe which only lives on steep slopes, and for whose members it is a matter of survival to never climb up or down the fall-line, and to avoid avalanches which come down the fall-line, then an indicative and a predicative word pair for steep planes could also very well occur, and a corresponding experiment could succeed. We, who have grown up in rooms and cities with horizontal floors and vertical walls at right angles, should also not assume that natural tribes which live in jungles, for instance, and only produce round huts and round receptacles, form the same system of distinctions as we do concerning front/back in interaction with left/right. Nor can we predict how they would answer in Uexküll's experiment.

The fascination which comes from Uexküll's text, and the deep truths which are to be found there against assumptions concerning the animal world that are much too naive, must not distract us away from those premises which Uexküll, for his part, accepts unreflected. He ignores, for example, the cultural load of his own language. Therefore, he misses the reference to something that is not present in every single individual merely as a natural, organic endowment, but also a system of linguistic distinctions, whose reliability lies in its communicative usefulness, hence, in a high degree of intersubjective reenactability. Evidently the three-dimensionality of space is the *premise* and not the result of the Uexküllian *theory*. One might argue that the triad of the colonnades, hence the triad of spatial dimensions, are the natural endowment of humans, quite apart from this theory. But this is false. Why are the three colonnades made responsible for the spatial dimensions? After all, why would we find the thesis absurd that space has five dimensions because every hand and foot has five extensions? Reference to the colonnades of the organs of balance is first made possible by the system of three mutually perpendicular planes, and elementary experiences with motion, self-motion and the motion of others, both as changes of position and as changes of attitude. Otherwise we would not even have the means to appropriately describe this organ in its spatial structure, or even in its functions.

Uexküll is very brief when it comes to the function of the colonnade. As a biologist he assumes that this is already known, or that it can be read about in the appropriate books. But here, too, an important problem remains open. It is more than questionable whether the colonnades of a fish can accomplish what Uexküll requires of them. Because of the gravitational field of the earth, the vertical direction is distinguished. But the other two react "inertially," that is, only against accelerations. These can just as well be "attitude accelerations" (for humans, say, turning the head) as position accelerations. But they do not react when a fish floats with a river current, without any swimming motions of its own. Then the organ of balance would also not have the homing effect which Uexküll attributes to it. It is a representation shortened at least to incomprehensibility when the explanatory chain is constructed thus: The homing effect is important for the survival of the animal. It is produced by the three colonnades. Thus, the animal finds itself at home in a three-dimensional space.

THE EVOLUTION OF THREE-DIMENSIONAL KNOWLEDGE

"Evolutionary Epistemology" (EE) comes at the end of all the approaches to the explanation of the three-dimensionality of space which we will discuss in the first part of this book, not just because it is historically the most recent attempt, but also because it is the attempt with the strongest presuppositions. EE refers not only to a biological theory about the origin of organisms in their modern form, but also to mathematical and physical theories about the world, to empirical physiological and psychological theories about perceptional and cognitive accomplishments of living beings, and even to scientific analyses of the modern natural sciences. At least in the form in which it is represented by its most important proponent, Gerhard Vollmer, it brings together a synthesis of practically all of natural science's states of knowledge, in order to then go beyond the older philosophical epistemologies, like those of Classical Empiricism (Locke, Hume), for example, or Transcendental Idealism (Kant).

In spite of this wealth of hypotheses and relations, the fundamental idea of EE is temptingly simple: When it is true, as modern biology teaches, that the modern endowment of all organisms is the result of a process of adaption to the environment throughout natural history, then this is also true for the organic endowments which serve us in deriving knowledge. In reference to perception (as a sub-realm of knowledge derivation) this means that we can perceive the world because our perceptual apparatus (sensory organs and nervous system) has adapted to the perceived object in a long struggle for survival throughout natural history.

Knowledge of this sort is defined in EE as an accomplishment of an organic cognitive apparatus. The distinction between knowledge and non-knowledge consistently results from the adaptive character of this cognitive apparatus to the given world structures. This adaption resulted from the interplay of mutation and selection according to the Darwinian theory of evolution, and its further modern developments (above all, in system theory). Formulated somewhat sloppily, our knowledge (of the world) is applicable (to the world), because otherwise we would not have survived. A general discussion of this theory cannot take place here. We are only interested in the numerous and sufficiently detailed comments on the three-dimensionality of space. Here again, Gerhard Vollmer is the clearest and most informative author.

In spite of unavoidable simplifications, Vollmer's list of epistemological alternatives concerning the three-dimensionality of physical space is worthwhile reading. The following is an adapted, shortened version of his list of the traditional answers to the question "Why is the world (physical space) three-dimensional?"[44]

1. Occasionalist (Geulinox, Malebranche): "Every time that I observe something three-dimensional, God wakes a three-dimensional image in me."
2. Pre-established (Leibniz): From the beginning, God created both substances, body and spirit, body and soul, such that they remain in harmony for all time; the clock metaphor): "Objects and intuition have always been three-dimensional."
3. Empirical (Locke, Hume): "Intuitive space is three-dimensional because the world around us is."
4. A priori (Kant, Eddington): "The empirical world is three-dimensional because our a priori form of spatial intuition is three-dimensional."
5. Transcendental-linguistic (Wittgenstein, Whorf): "Empirical space is three-dimensional because language attributes three-dimensions to it."
6. Conventionalistic (Poincaré): "The world is three-dimensional because we have agreed to describe it this way."
7. Economic (Spencer, Mach): "The empirical world is described three-dimensionally because this description is the most economical."

   Vollmer places the view he shares at the end of this impressive list.
8. Evolutionary (Lorenz: "Some cognitive categories were developed in adaption to reality, that is, were acquired phylogenetically. For the individual (ontogenetically) they are inbred."): "The empirical world is three-dimensional because our spatial intuition developed phylogenetically in adaption to a three-dimensional world."

---

44 G. Vollmer (1975), p. 55

The thesis that is formulated in Vollmer's first book *Evolutionäre Erkenntnistheorie*[45] concerning how our brain can "three-dimensionally interpret" the two-dimensional retinal images of objects in the outer world, is only discussed as a question there. It is then dealt with more extensively in his book *Was können wir wissen? Die Natur der Erkenntnis*(1985) (What Can We Know? The Nature of Knowledge), which was inspired by a very lively discussion.

The book begins with a preface from Konrad Lorenz in which he says:

> What the Uexküllian theory of environment gave me personally was the realization that many realities which appear completely comprehensible to us humans, are insurmountably transcendent for more simply constructed organisms. Deeply convinced of the organic nature of my own cognitive apparatus, it therefore seemed to me a wanton arrogance to see the boundaries of our human cognitive capabilities as absolute, and to want to pronounce a final "to this point and no farther." Slipper animalcule live in a world of only one dimension, belly-haired infusoria live in a world with two, and all spatially-oriented living creatures obviously live in the same three-dimensional space as we ourselves; who would dare to assert that this must be the last possible orientation in space? Although we certainly know of no animal that, like the spaceships in science fiction, could disappear out of our world by passage into another dimension.[46]

Let us read further in Vollmer, assuming for the moment, as the authors Lorenz and Vollmer do, that this statement of Lorenz's will be explicated by Vollmer's presentation in his book, or is at least be not incompatible with it. (In fact, they are not compatible, so that the quoted passage by Lorenz will have to be considered for itself.)

In Chapter Nine, *"Das projektive Modell"* Vollmer comes to the three-dimensionality of space, with a purpose which is of central significance for the entire undertaking of EE. This subject is not only an example of a special cognition, which is then to be interpreted evolutionarily. Rather, it is typical of his entire epistemology, which is called "projective": environmental cognitions are projections of world structures upon our (organic) cognitive apparatus. That is, from the wealth of a world "independent of consciousness, structured by laws, and coherent" something is "partially"[47] mapped through projection, under conditions of simplification and information loss.

---

45 G. Vollmer: Evolutionäre Erkenntnistheorie. [1]1975

46 K. Lorenz in: Ibid, p. XIV

47 G. Vollmer (1985), p. 28

The example of spatial dimensions is dealt with through an analogy argument: one imagines flat, hence two-dimensional, living beings. If these beings formed a science, that is, a measuring natural science, then they could come to the empirical-scientific conclusion that they lived on a sphere, for instance, through the determination that the sum of the angles in a triangle is larger than two right angles. This would not, of course, be intuitively imaginable for the flat creatures (as it is for us), but it would have exactly the same mathematical properties as it does for us, as three-dimensional beings. Such flat creatures could proceed from a two-dimensional figure (of the sort that already occurred in the discussion of the language-analysis interpretation; Cf. Part I, Chapter Four), interpret this as the projection of a three-dimensional cube, and draft a mathematical theory for a space of a "higher" dimension number, namely, higher than the dimension number of the everyday, flat empirical space. This would then have a confirmed application in an empirical physics (the comparison is drawn with the discovery of the spherical shape of the Earth, which was held to be a disc in some cultures), as in the above-mentioned determination of the curvature of the surface.

By analogy, we humans can mathematically design a four-dimensional "cube" (Vollmer: "hypercube" with 16 corners, 32 edges, 24 surfaces and 6 bounded spaces[48]), and apply the theoretical knowledge thus gained within empirical physics. For Vollmer the point of this analogy is to bring *science* into play. The circularity reproof which has been made against EE is that, for the thesis of the adaption of cognitive structures to world structures, it must already be known how the world is. Vollmer believes to be shielded against this reproof by the fact that he accepts "theoretical cognition" (that is, results of physics): "This is an extremely important point which is overlooked by most critics, and even representatives of Evolutionary Epistemology. It will even resolve the apparent circularity."[49] We will come back to this during the critique of the Vollmerian approach.

Even here, in his reply to the critics, three-dimensionality is picked up once again. The end result is:

> The fact that the world "really" has three dimensions is confirmed by all relevant theories of physics. There is not the slightest clue as to a deviant dimensionality, no contradiction, and no better success of a four or n-dimensional rival (although such rivals have repeatedly been suggested and tried) ... There are even proofs that under several very general and empirically well-confirmed conditions

48 Ibid, p. 29
49 G. Vollmer (1985), p. 254

> (resonance-free propagation of waves, closed planetary orbits, possibility of living systems) the physical, macroscopic space *must* be three-dimensional. If it were not, then some things which are found in our world would not be possible, or would be different than they are.[50]

In other words, Vollmer admits that EE would be circular, and therefore untenable (at any rate, I have found no passage that would characterize a circular structure of this epistemology as harmless), if it were only supported purely by everyday or perceptual knowledge of how the world is, in order to prove the adaption of our everyday and perceptual cognition. According to Vollmer, however, EE is saved by the fact that the *natural sciences* tell us how the world is. In reference to the dimension number of space, physics answers "three," which is why it is no wonder that our cognitive apparatus (sensory organs and nervous system) prefers three-dimensionality, in "partial isomorphy" to real space and conceptual space (which is formed by a "fusing" of, among others, the visual, aural and tactile spaces[51]). Or, more briefly, because empirical space is three-dimensional, as the natural sciences have proven, the human race has developed in such a way that humans also perceive or conceive space three-dimensionally.

The first item of importance for a critical evaluation of the contribution of EE to the dimension problem of space, is the fact that not one definition of "dimension," of the dimension number, or of "space" is brought forward. Rather, ordinary language usage is accepted, according to which a point is zero-dimensional, a line one-dimensional, a surface two-dimensional and a body (e.g., sphere, cube) is three-dimensional. This applies to Vollmer and Lorenz alike.

For Vollmer spatial three-dimensionality is an unimpeachable fact of natural science, from an analytical perspective. The examples mentioned (the echo-free propagation of waves and the stability of planetary orbits) are precisely the kind which were referred to above, mainly in the proposal of Paul Ehrenfest; and the same objections apply. The fact that one can derive a distinction of three-dimensionality from completed, acknowledged physical theories (which is not at all disputed here) says nothing about its epistemological status, and nothing at all about the fact that this knowledge is experimentally and theoretically supported. Because, to the extent that it still wants to distinguish itself as a real science, as opposed to the ideal science of mathematics, physics remains bound to technical procedures of control, of observation, of experiments and of measurement, regardless of whatever lofty abstract theories it might aspire to. And the language that still rules is that of the

50 Ibid, p. 62
51 Ibid, p. 49

engineer and the mechanic who build the necessary instruments, and of the experimental physicists. These specialists do not allow themselves to be dissuaded by any theoretical considerations from describing the form, magnitude and positions of bodies in the normal, three-dimensional way.

What Vollmer completely overlooks in his rescue of EE against the reproach of circularity is that his own Archimedean point, namely the empirical knowledge of the natural sciences itself, has an input of dimension theory and practice: to wit, the technical practice of interaction with bodies (language-free, as well as conceptual). Bodies are artificially worked, or fabricated, into the apparatus of an experimental science. In the process, the actions of technicians who produce (and use) these instruments are formulated into rules, all of which are described with Euclidean geometry. Vollmer's assertions are, *de facto*, an assertion of the adaption of our (spatial) cognitive apparatus to a world technically outfitted by humans - but one might very well doubt that this is, or could be, the intent of EE. Even additional or helping hypotheses, which state that the possibility or success of this technical practice shows the factual three-dimensionality of our environment, do not help us further here. Before such arguments could be of use, one would first have to clarify the extent to which this judgment about the possibility or success of technical practice is dependent on experience and stipulation.

Vollmer's proposals cannot be criticized because the physics and evolutionary biology that are presented in them are insufficient or false. The decisive weakness lies rather in the realm of science theory. His actual mistake lies in ignoring a central question: What is it that distinguishes the results of physics as knowledge? Have some decisions already been made at the foundations of physics, which Vollmer has mistaken for its *results*? Since we have seen that this flaw occurs in his defensive strategy against the reproach of circularity, we must ask ourselves if this defense has, perhaps, also failed; and with it, the entire EE. But this question no longer fits thematically in this book.

At least now it has been implicitly shown that Vollmer represents a different perspective *in puncto* three-dimensionality of space than Konrad Lorenz. No reference is made by Lorenz to the results of physics, but only to a hierarchy of the realms of living beings. In his impressive, although critically vulnerable, book *Die Rückseite des Spiegels* (The Other Side of the Mirror), Lorenz uses examples of living beings to present a systematic construction of increasingly complex cognitive accomplishments, which is, in a very decisive regard, primarily a theory about accomplishments of spatial orientation (with the goals of food, mating and flight).

The amoeba as a spatially unstructured living creature, without the distinction of a body axis, reacts unspecifically in space. However, it must at least distinguish between backing away from a stimulus and flowing toward it, in order to enclose food, for instance. The slipper animalcule already exhibits a body axis and motion preferences (through inhomogeneity and the connected reaction potentials to environmental stimuli) even if, after backing away from a rejected stimulus, no particular direction of motion can be distinguished. Rather, after rotating to any given angle, a new attempt at motion is made. In a disadvantageous case, this can lead back to the same negatively valued object. Correspondingly, belly-haired infusoria, with a two-dimensional motion repertoire, are discussed, and finally, animals with a three-dimensional reaction repertoire (retrospectively classified higher) are considered.

As impressive as this theory is (especially when it turns to the higher accomplishments and distinguishes, for example, between the mere playback of inbred programs in animal behavior and learned answers to situations), it is still not clear which principle is responsible for this neat order. Which system is referred to here? Is a development theory of the historical sort intended? At least accomplishments of *modern* living beings are being considered. It is a question which must be answered biologically, and not by philosophers, whether living creatures like slipper animalcules are biological prototypes which have been left over (to which the ancestors of man also belong), or descendants which developed in separate branches of the evolutionary tree; the one to a simple back and forth, the other to Euclidean geometry.

Even if this question remains open here, at least one question can be clearly answered. Lorenz most certainly brings a system of conceptual distinctions to bear on his observations and systematization of the animal realm and its behavioral forms, in which "dimensions" are already present as motion directions or as symmetry levels for the bodily construction. So the question of whether or not a very anthropomorphic conceptualization (namely that of the architect, carpenter and engineer) is brought to bear on nature as a judgment pattern, in such a way that the answers are already structured, and perhaps even given, must clearly be answered "yes."

Besides the cognitive aspect, however, there is also an emotive one to be emphasized. One may assume that an admirer and friend of nature like Lorenz, with an almost religious reverence, is sincere when he shows modesty. He asserts that it is "wanton arrogance ... to see the boundaries of human cognitive capabilities as absolute." And it is, of course, rhetorical when he asks, "who would dare to assert" that the specifically human three-dimensional view of space "must be the last possible orientation in space."[52] To imagine oneself to be modest, and to actually be modest are, however, two different things. In order for Lorenz to be able to reduce humans (and himself) so far that he sees them as one animal among many, at the mercy of their unfortunate rationality, and subject to natural history as the final arbitrator, he had to have taken up an empirically very immodest observation post outside of the specifically human cognitive boundaries. To wit, an unhistorical standpoint which views natural history as a whole, free from the disturbances of all human rationality standards (like that of his own theory).

But how are the boundaries of the cognitive accomplishments of his own theory to be viewed in terms of Lorenzian relativity; what happens when one applies Lorenz's theory to itself? One should consider this question in regard to the various reserves of knowledge which are commonly held to be important for human life, from personal childhood memories to mathematics, from the section of the environment seen at a given moment, clear to the theory of relativity, from the child's accomplishment of distinguishing between honesty and lies, up to the Lorenzian theory itself. All of this is supposed to be comprehensible under the single aspect of considering cognition as the accomplishment of an organ apparatus. Without considering the historical and cognitive reasons for picking out particularly this aspect, and without considering the linguistic character of this very description, it is undoubtedly a much neater theory. But what kind of validity can an epistemological theory possess that ignores the social and cultural accomplishment of intersubjectively (indeed, transsubjectively) verifiable sentences, capable of consensus, or that proceeds without reference to human goals, to rules, principles, and stipulations? An epistemological claim is made here which, even though framed in the secularized form of a reference to natural laws, borders on godlike. It simply ignores all human achievements that came *before* biology and evolution theory. It is hard to imagine an epistemology that could be less modest.

SUMMARY

Attempts to trace the three-dimensionality of space back to sensory perception and to the organic endowment of the body have been presented in two very different types of approaches. The one type, represented here by Poincaré, proceeds from a non geometrical, topological definition of "dimension" for mathematical continua and transfers this definition to the physical continua of sensory qualities.

52 Cf. Note 46

Such theories must then generally show the interaction of various senses for the construction of a space. The other type, represented here by Uexküll, Vollmer and Lorenz, makes use (without theoretical scruples) of the normal conception that three mutually perpendicular planes set up a three-dimensional space. This approach then gives various reasons from the realms of biology and physics, as to why space is three-dimensional in this sense for us humans. For Uexküll, it is a matter of functional potential, i.e., our bodies' potential for movement, and of the organs oriented to motion control, especially the organ of balance. According to Vollmer, it is shown in the historical development of our cognitive apparatus in adaption to a three-dimensional world, whose factual spatial structure we learn from physics. Finally, for Lorenz, it is the position that organisms, including humans, assume in a system of organisms with various degrees of spatial differentiation.

In spite of this clearly separated classification, these approaches are very similar, particularly those of Poincaré and Vollmer. Regardless of Poincaré's authorship of the position of conventionalism in science theory (Vollmer must have oriented himself more on Poincaré's label than on his actual assertions, when he counts Poincaré's view of three-dimensionality as one of the alternatives to his own), both authors agree that spatial three-dimensionality is distinguished by physics. Even Poincaré, who thinks thereby of invariance properties of physical laws, stresses numerous times the empirical or experimental character of this distinction. (On the other hand, not even Vollmer would want to dispute that physics is not possible without any conventional stipulations at all, even if just in terminology.) Poincaré explicitly adds that our knowledge of natural science concerning the construction and function of our perceptual organs was likewise achieved by the relevant sciences through the application of the normal, three-dimensional spatial understanding. Vollmer also implies this in his references to perceptual psychology and physiology.

In other words, the three-dimensional and Euclidean technical practice of gaining experience in physics, as well as in physiology and psychology, is the self-evident, but unrecognized foundation for Poincaré's, as well as Vollmer's, proposals (regardless of Poincaré's topological definition of dimension). For this reason, an appropriate criticism of both proposals is that an understanding of science which does not sufficiently reflect the foundation of physics, and which ignores the purposeful artificial-technical production of the objects of science by the rationally-acting researcher, ultimately leads to an insufficient result. As the logical consequence of theories, one only discovers those propositions which have served as the technical and empirical foundation of these theories from the beginning.

Even Uexküll's and Lorenz's explanations for three-dimensionality exhibit great similarities. In the end, both make use of the motor accomplishments of organisms, and their control by the corresponding perceptual organs (oriented to internal or external processes). Neither holds "space" to be a given, the same for all living creatures, but sees it rather as being constructed by each organic individual with its motor and sensory accomplishments. To this end, both make use of the usual geometrical means for descriptions, as if the legitimacy of this were self-evident. And in their texts, the question of whether one would arrive at the same conclusions if classical naivety and forgetfulness in regard to one's own linguistic means were not the ruling principles (as they were in the classical physics of Galileo's and Newton's day), is not only left unanswered, it does not even once appear on the horizon.

CHAPTER SIX

Euclid's Heritage: A Review of the History of the Problem

When so many prominent scientists and philosophers address a problem, which is obviously the case with the dimension problem, as can be seen in the preceding presentation (which is not even intended to be complete, but merely representative), one of two assumptions suggest themselves. Either it is a very fundamental problem concerning an important principle, or it deals with an illusionary problem which is difficult to recognize. After all, even illusionary problems can be extensively disputed. Therefore, we want to review the discussed approaches to spatial three-dimensionality with the following question in mind: can points be found to support the first assumption that a clearly specifiable problem, hence a philosophical task, exists?

What is most conspicuous is that not one of our authors had any difficulties with the linguistic means. From ancient times, when the three extensions of the body (or spatial geometrical figures like the sphere, and the five Platonic polyhedra) were first accepted, to the beginning of modern science when Galileo assumes three planes which are pairwise perpendicular to one another, as if they were self-evident, from the philosophical classics up to the modern natural scientists and philosophers of science, it always seems immediately clear what is meant by the three spatial dimensions. And even where explicit definitions for "dimension" or "dimension number" are given, by Clifford, Carnap, Poincaré and Brouwer, for example, these only serve to sink these definitions more deeply into the construction of mathematics. It is not a measuring or projective geometry that first allows a definition of dimension; on the contrary, set-theory, or the continuity of motion, or even the most general properties possible for the classification of spatial forms, already make such a definition possible. In other words, even here one proceeds from general, geometrical (or better, form-theoretical) understanding, and attempts to demonstrate that there are more general definitions which succeed with even weaker presuppositions.

But as far as we can see, no explanation has been found for the far-reaching correlation that obviously exists in linguistic distinctions and which crosses over historical and cultural boundaries. It is self-evident that the analytical arguments of Carnap, Poincaré, Ehrenfest, Vollmer and others (which state that when the same natural laws are valid everywhere, then the structure of space in natural law must also be spoken of the same everywhere) do not explain this fact. Highly theoretical constructions, which still mostly remain unknown, are needed to even assert a connection between physical laws and the three-dimensionality of space.

The obvious correspondence in linguistics, which has shown itself in the various approaches, must have its basis somewhere. Even the construction and function of our sensory organs do not come into question as an explanation, as we have seen. Even there, it is not the simple utilization of sensory perception which "functions the same" for all humans, but its description in special empirical sciences like physiology and psychology, which first leads to a concept formation of space and its dimension number. However, these empirical sciences already make explicit use of particular concepts of space and dimension.

As candidates for the explanation of this conspicuous phenomenon in the history of the mind (one need only compare how very differently other realms of epistemology have been put into language, where every philosophy immediately offers its own terminological system of distinctions) only the purely spatial approaches come into question; the others already make extensive use of the distinctions made there. This is even true for kinematic approaches (which make reference to motion). The basic vocabulary of Euclidean geometry, the words "point," "line," and "surface," as well as relational expressions such as "to cut," "to incide," or "to bound," form the set of linguistic instruments which is applied everywhere. There is no alternative anywhere for precisely this trio (point, line, surface) and a relational expression (either to cut, or to incide, or to bound). This is the first conspicuous item in the review of the history of the problem, which stands in a close logical relationship to a second one.

None of the approaches even considers the possibility that it is not bodies - the physical things, like stones and living beings, instruments and buildings - which are valid as prototypes for three-dimensional spatial objects. For some philosophies, like the Cartesian, where the world of bodies as *res extensa* is set against the world of thought, the suspicion might arise that three-dimensionality and corporeality are *definitionally* bound to each other, so that the representative character of the body for spatiality would not be very surprising. But even more likely is that this quasi-definitional tie is held to be unproblematic in Cartesian philosophy because volume, space-filling and corporeality had already been equated for thousands of years. One need only think of the demonstration of the corporeality of air in ancient Greece. Already back in times that our written tradition barely touches, bodies were described by shape, and compared with others in position and magnitude, with a language that contained the basic vocabulary of geometry. Craftsmanship and art made use of this language (and still do today) without the epistemological finesse which the comparison of a physical world of bodies and an ideal, mathematical world of thought demands. Regardless of what has been presented from the modern positions of science analysis about the relationship between geometry and reality, in practice, the architect and the pilot, the physicist and the psychologist, indeed, even the crafting mathematician, all communicate without any problem using the words of geometry's basic vocabulary about bodies.

It can be seen as an anthropological cultural invariant that people work on bodies, produce instruments and dwellings and, in order to hand these artifacts down, develop a language to teach these skills. If three-dimensionality was bound to the trio of the words "point," "line" and "surface" (as boundaries or intersections), which form the basic vocabulary for talking about bodies, their form, magnitude and position, then the high cultural invariance in the view of the three-dimensionality of space, that is, the spatial properties of the body, could be understood.

Of course, then the pressing question immediately arises: what distinguishes these three from a much larger number of geometrical expressions as primary or basic terms. The fact that the definitions in Euclid's first book are ordered as they are, already demonstrates systematic scientific considerations of a high state of development. Not only the sequence of point, line, and surface count to this order, but also the order of beginning with the definitions of the general concepts (line, surface), and then proceeding to the definitions of the subspecies of these (straight line, plane surface). These systematic scientific considerations must be taken into account here, because we wish to explain the high-percentage correspondence in the basic view which informs spatial speech about bodies, particularly among scientists and philosophers of practically every orientation. In other words, if only the definitional-theoretical maxims of the historical Euclid or an ancient school of mathematicians were to account for the distinction of the conceptual triplet, without these maxims corresponding to anything equivalent for the layman or craftsman, then the triad of spatial dimensions would still not be understood.

As already discussed above, mainly in the first chapter, in ancient geometry, talk of dividing and bounding was applied, without theoretical reservations, to the physical realm of real bodies as well as the abstract realm of ideal geometrical forms. This may leave the modern philosophical analyst unsatisfied - the epistemological question concerning how it is possible to apply the ideal geometry to real bodies remains open - but no reservations for the layman result from this. So we do not want to localize and infer the problem of three-dimensionality in the definitional - hence, conceptual - weaknesses of geometry as Euclid formulated it. The problem to be solved here is the problem of the application of geometry. The mere absence of definitions which fulfill the modern standards for the construction of an exact theory would provide too few reasons for the fascination with the problem of three-dimensionality. The actual (and unproblematic) use of the basic geometrical vocabulary in talking about bodies is far more effective than modern reservations of science theory can be, especially in retrospect to earlier authors.

It is another heritage of Euclid which raises the problem of three-dimensionality, at the same time that it obstructs the solution. *Geometry begins* for the Greeks *with planimetry*, that is plane geometry, or geometry in the surface. Geometry does not arrive at the spatial forms until Euclid's eleventh book. These are derived from plane forms through motion (rotation). One can recognize the reason for this from the etymology of the basic geometrical words for point and line. Geometry is a theory which idealizes the *practice of drawing*. Drawings are planar. An important decision for geometry, which is still in effect today, was tacitly made in the choice of words (even for the ancient geometricians, the words for point and line were already part of general language usage before definitions were given for them), in the reference to the drawing and painting arts, and certainly also in the demonstration of planimetrical states of affairs in the sand or on wax tablets. The fact that the drawing surface was, itself, the surface of a body, was not the object of geometrical considerations. Thus, absolutely no use was made of important body properties. Indeed, who would draw a geometrical figure in a fluid or on a Swiss cheese? A beginning - indeed, the planar foundation of geometry - was made with the tacit choice of words, and the reference to drawing practice, but the indispensable corporeal properties of the drawing instruments were not taken into account in this.

Consequently, geometry arrived at a drawing practice in which the line is represented by a stroke, and the point by a small spot of color, or by two intersecting strokes. This is a principally different starting position than that of Clifford's proposal, where the line is defined as the boundary between two neighboring spots of color, which itself has no color and no planar extension, and this is analogous for the point. In other words, the *reference to drawing practice*, and the historically fateful decision to *begin geometry as planimetry*, have obscured the boundary character of the point and line (and that of the surface as well). (We have mentioned that Aristotle explicitly stressed that being the boundary of a line is not the essence, i.e., it is not an exhaustive definition, of a point; an indispensable definer is its indivisibility.)

As a result, through philosophical arguments (as in connection with Aristotle's repudiation of Zeno's paradoxes) the point, as the indivisible unit, became the basic concept of all geometry and, in the end, was only treated by geometricians and mathematicians in the synthetic generation of forms of higher dimensions - the common practice even today. It is clear that three-dimensional sets of points can thus be illustrated with a physical body, just as surfaces and lines and points can be illustrated as its surfaces, edges and corners, or as corresponding figures painted onto it. But it is also clear that, without a structuring interaction with bodies, geometry (including its historically factual start in drawing practice) would have neither a material foundation, nor illustrations for mathematical entities (sensually perceivable and technically producible), no matter how abstract, idealized, conceived, defined or interpreted.

The dimension problem and the endless philosophical reflections on "space" also have their origins in Euclid and his decision to begin geometry as planimetry. If it had begun as stereometry, as interaction with bodies (meaning primarily those bodies which are used in geometry lessons as objects for illustration, and of course the drawing instruments as well: the drawing board, the rulers, the compasses and pencils), then it could have also been seen as an ideal science about the spatial properties of real bodies.

In this case, an upper boundary for sets of points must not be given as an argument for the triad, but reasons must be given as to why a three-dimensional language or terminology is suited for the interaction with the objects of demonstration.

In the following systematic chapters, reasons will be given as to *why the concepts of boundary, or of intersection, alone are not sufficient* to this end: the cutting of bodies, and the resulting artificial production of surfaces, edges and corners, does not end with corners; a body can be further cut through corners. Instead, certain *forms of cuts* must be taken into consideration, just as was finally the case in all of the approaches discussed above, including the "topological" one. And this is the third conspicuous item which the history of the problem exhibits: The triplet of pairwise perpendicular planes is ultimately the indication par excellence for three-dimensionality. The fact that the word "*dimensio*" refers to measurement, i.e., the comparison of magnitude, plays no more of a role here than the long-known and trivial fact that the localization of bodies in space is also clearly possible with coordinate systems of oblique angles.

Most ordinary languages distinguish between length, breadth and height, but do not know a fourth concept which belongs to this group. We fix directions in reference to our bodies using a triplet of planes. We can densely stack square or cube-shaped objects, or can rationally build them into a wall. These are all *distinctions of form geometry in practical living*, which get by, for their part, with a *triplet* of surface, line and point for good, explicitly nameable reasons. It is precisely the *plane* surface cuts and, as their cuts, the *straight* edges, and then the *right angles* between edges which deliver the triad. One can, indeed, make yet a further cut through the corner of a cube, but not one which stands perpendicular to all the rest - which will be shown in Part II.

The initial question, about whether this is a philosophical problem or only an illusionary one, can now finally be answered. The epistemological task is twofold, namely:

1. to make an unknown mistake obvious through analysis (the mistake, adopted along with Euclid's fundamental decision for planimetry, was to begin geometry with a sequence which was critically disastrous for the construction of the theory, because it ignored the world of bodies and all of its relevant properties), and
2. to show that without this mistake the old riddle of three-dimensionality can be solved, not in the fields of topology or formal analytical geometry, but through the form-theoretical foundation of geometry - the technical and linguistic practice for unequivocally reproducible cuts through real bodies.

PART TWO

Space Is Three-Dimensional: What Does It Mean, and Why Is It True?

Introduction

The thesis that this book attempts to clarify and prove, namely that "space is three-dimensional," is formulated in familiar words and is, indeed, itself a familiar sentence. It therefore suggests an immediate comprehensibility, just as it presumably stimulates a sort of consent reflex, as it were, in every competent speaker of the English language.

The preceding chapters have dealt almost exclusively with scientific or philosophical theories. But the great intuitive certainty with which the layman is convinced of the three-dimensionality of space should not be forgotten. An elucidation and foundation of the thesis of three-dimensionality must do justice to this as well, i.e., the following explanation must deliver evidence of its relevance for everyday understanding, as well as for physics or philosophy.

This goal recommends special caution, particularly at the beginning of the discussion, because primary decisions which obstruct a close reference of later results to the clear intuitive conviction of everyday understanding, are made all too easily. Already the first sentence of this chapter, for instance, contains a type of preliminary decision, the justification of which one might reasonably question: is it really best to raise, on one hand, a problem of meaning and, on the other hand, a problem of truth - to ask what the words "space" and "three-dimensional" *mean*, in order to ask, as a second step, whether the thesis is *true*? (A common alternative, often chosen, would be not to separate conceptual clarification and truth content but, to proceed from the assumption that one already knows what is true, and to attribute the corresponding meaning to the words which occur.)

The everyday speaker and dimension-theory layman will probably *not* immediately feel himself required to distinguish between aspects of meaning and aspects of truth. Even in response to an inquiry about meaning and truth, he would presumably not begin by bringing forward a few definitions, with whose help he could then, in the second step, show truth - whether logical or through reference to certain experiences. He would rather answer the question with reference to some examples, which illustrate for him the meaning of the theses and, at the same time, demonstrate their truth. Perhaps the layman would make reference to the triad of directions relative to our body (above/below, left/right, front/back) or to the length, breadth and height of any given room or brick, and perhaps, more generally, that of any body. Such examples are familiar to everyone and are completely accepted. On nearer observation, they have the same diffuse character as the thesis from dimensionality itself. Because the layman would, indeed, neither be inclined to accept them on the basis of mere linguistic rules or conventions (what are other linguistic rules or conventions supposed to look like then?), nor would he believe he knew them based on experience alone. After all, even the layman has an intuitive understanding of the fact that only something which could have been something different, or could have occurred differently can be dependent on experience. Even someone who has never heard of the difference between empirical and a priori statements knows that today's weather can only be known from experience, while he would never regard a simple mathematical sentence like 3 + 5 = 8 as empirically dependent (the consequence being that on the next day one must calculate anew to see if the sum of three and five is still eight). Correspondingly, it cannot be that, tomorrow, we would want to distinguish a new direction relative to our body, or to add a new extension to our rooms, or exclude an old one. The examples, like the general thesis itself, are, as is often said, "necessarily" true.

These are, then, the three aspects which are accessible to the layman:

1. the fact that we cannot even give illustrative examples for the three-dimensionality of space without certain linguistic distinctions,
2. the fact that, in this, we always make reference to the outer world, in the form of our own spatial orientation or as the spatial properties of bodies; and finally,
3. that it seems to us that it could not be any other way. In the following chapter we must do justice to all three of these aspects every bit as much as to the acknowledged theorems of mathematics and physics. (A healthy skepticism is only appropriate regarding philosophers - even if they are actually mathematicians and physicists by profession - in the face of their previous lack of success in this matter.)

With this goal in mind, the second part of this book does not begin with a leap into the middle of a theory or a solution. Rather, it first tries to work out an approach to the problem. It asks what we are even talking about when we speak of three-dimensionality, what we might possibly accept as an evident answer, and which criteria we can bring to bear. Also, what can be intended when we speak of "space." What are we searching for, i.e., what kind of knowledge, in reference to this "object space"?

CHAPTER SEVEN

Knowledge about Space

"SPACE" AS OBJECT

In the thesis "space is three-dimensional" we treat space linguistically exactly as we treat the moon or the Eiffel Tower in the sentences "the moon is spherical" and "the Eiffel Tower stands in Paris." The grammatical correctness of the three sentences and our familiarity in ordinary language with such sentences suggests to us that space is an object in the same way that the moon is a natural object and the Eiffel Tower is an artificial object. This means that the *linguistic form* of the thesis creates the appearance of substantiality, as if "there were" a proper thing about which we can make statements concerning when it is found, where, and what its other properties are.

This linguistically suggested similarity between space (where the lack of the indefinite article signifies that there is precisely one) and the things moon and Eiffel Tower breaks down rather quickly, however, without having to bring large caliber philosophical guns to bear. At the same time, every layman is certain that, if he had enough time, interest and perhaps also the proper aids, he could practically find as many new statements, demonstrable as true or valid, about the moon or the Eiffel Tower as he wished. With space, in contrast, the linguistic means for this would quickly be exhausted. Words occur such as relative and absolute, empty, finite and infinite, perhaps even words which come from popular presentations from physics, such as curved, flat and a few similar predicators, but then the list of possible statements would be exhausted. The difference between space and the two other objects would be even clearer if we asked about the manner in which we arrive at new statements. In the case of the moon or the Eiffel Tower, we would, of course, examine the object more precisely, prepare pictures, photos and drawings, try to get closer to it, perhaps carry out measurements, take material samples and much more. In case of the object "space," none of this would be possible.

Perhaps one might now object, "of course we can find out something about space through measurements and scientific observations or experiments." But this objection overlooks the fact that the physicist, who is most qualified for the empirical research into "space," does *not* treat "space" like "the moon." Instead, in the case of space, he carries out investigations into the structure of fields, into the motion of bodies, into the spread of light rays, etc.; and he unites these all under the title of "research into space." In other words, it is precisely these physical theories about space which provide an indication that to talk about space is an abbreviated form for making *spatial statements* about *other* things like bodies, forces, motions, light rays, etc. In short, to speak of *space* means to make spatial statements.

For the layman, inexperienced in the philosophy of language and definition theory, this might appear to be sleight of hand: one frees oneself from the mysterious metaphysical object space by freeing oneself from the *word* "space," and one does this by using the word "spatial" as a predicator for statements or words.

The suspicion that one has not gained anything by replacing the word space with the word spatial is, however, easily dispelled. "Spatial" is supposed to be applied in the classification of other words, i.e., "spatial" is a meta-language word for the labelling of other words. With nothing more than competence in ordinary language, we can effortlessly put together a list of the words we use for spatial statements. When we say, for instance, "a piece of wood in the form of a cube, with 5 cm. long edges lies on the table, and when we tip the table this cube begins to slide in a straight line to the edge of the table," then we have an example sentence in which, except for the words "piece of wood," "table" and "begins," only spatial words occur. The language competence of everyday life likewise suffices to say that with spatial words we speak about form and magnitude, (relative) position and motion of bodies (to the extent that temporal determinations are not being made), and perhaps from other objects as well, like spatial statements about a storm. Hence, the danger of a definitional circle does not arise when we use the word "spatial" instead of "space." That means, we do not have to speak anew about space or any of its "properties" in order to determine an application of the word "spatial." We simply write down a representative list of words and propose to *call* them all "spatial" words.

But it still might appear unsatisfactory to the layman that the honorable old philosophical problem of space supposedly disappears because of the definitional deduction of one word from another. This perspective, however, itself contains an error. No honorable philosophical problems are supposed to be made to simply disappear through a definitional trick. On the other hand, it would be just as fully unjustified to assume that a serious problem is hidden within every incomprehensible formulation in the philosophical tradition. Until now only the first step in the elucidation of the question regarding what it means to speak of space has been undertaken. All questions, theses, opinions, errors and so on which, according to the traditional characterization, concern space, must now be identified as questions, opinions, theses, errors and so on, regarding spatial statements. One can see, however, that no contextual or thematic limitations or exclusions have yet been generated.

Here it is only a matter of clarifying that the question about the meaning and truth of the thesis of the three-dimensionality of space must, henceforth, be posed as the question about which statements summarize or represent this thesis, and then, about what these statements mean and what their truth is based on. "Space" is not an individual object, but is, as a word, merely a thematical or contextual boundary of a realm of statements.

THE RELIABILITY OF KNOWLEDGE ABOUT SPACE

The intuitive and far-reaching certainty with which the layman assumes the three-dimensionality of space has already been mentioned. But the mere feeling of certainty, of course, neither imparts insights, nor is, in itself, a type of insight. Therefore, in what follows, we must consider which measures - possibly special measures for the theme of "the three-dimensionality of space" - can lead to knowledge and insight. "Reliability" shall be named here as the most provisional and general quality of knowledge and insights - which is, at first, nothing more than an allusion to an individual's attitude toward his own stock of knowledge or understanding. One relies on it, and one uses it, without continually having to reassure oneself about it.

Every person can confirm from his own life experience that one always relies, and must rely, upon private as well as public stocks of knowledge, hence upon highly specialized and, often, upon highly general stocks of knowledge, too. The distinction between private and general knowledge is not meant to be a value judgment. Nobody can live his life without a fund of private knowledge (e.g., about one's own biography, if nothing else), but likewise, nobody can get by without generally accepted knowledge.

Attempting to derive the general thesis of the three-dimensionality of space from the intuitive certainty of individuals (i.e., from the fact that one constantly relies on the forms of three-dimensionality, because they prove to be practical) is not a successful method. After all, these forms - our everyday examples of three-dimensionality - would still have to be described in language and be set in relationship to scientific doctrines, for instance. The individual's intuitive certainty does not allow us to ignore the linguistic form of knowledge about three-dimensionality. The mere fact of non-linguistic reliability contributes neither to a clarification, nor a foundation, of the thesis in question.

Therefore, reliability should, more precisely, not be understood as a mere factual interaction with stocks of knowledge, but as a *claim to validity*. Reliable knowledge should prove itself in as many cases as possible, even in unforeseeable ones. Normally one says that knowledge should be as *general* as possible. This corresponds to another usual practice (which, upon closer consideration, has proven itself well) of understanding this generality of knowledge in a dual sense. On the one hand, the applicability of knowledge should be as general as possible. On the other hand, the truth should be as general as possible. Hence, knowledge should apply to as many cases as possible, and should be tenable for as many people as possible.

The measures which recommend themselves for the more or less reliable derivation of these two generalities, belong to the framework of general standard knowledge in philosophy. To list and to briefly explain them in what follows serves the sole purpose of having them available later for the procedure which is actually taken up for the systematic solution of the dimension problem. In contrast, the next sections of this chapter are neither intended to be science theory nor epistemology; they are too specialized and too directed to the purpose of this book for that.

An important way to assure the general reliability of knowledge, was already demonstrated, in passing, in the little linguistic criticism regarding speech about "space." It is a matter of being able to express stocks of knowledge, or, more precisely, of the *linguistic explicitness* of knowledge. It has already been sufficiently explained that, in regard to the thesis of three-dimensionality, only knowledge which is linguistically expressed can be considered. What is meant by explicitness, beyond this, is that all relevant - in this case, of course, all "spatial" - linguistic means must be expressly known in their application. If they are only known to us through ordinary language usage, this explicitness must first be produced. Our ordinary language is, as a product of its own history, a collection of words and decisions whose historical origin is, in most cases, just as unknown to us as are the (we can only hope) once clear intentions of the originators to express distinctions. Unfortunately, the situation is not much better concerning the words of specialized scientific languages - in spite of the restriction of scientific testability, which arises with them. Here, too, one finds a historically-grown collection of terminologies which also cannot be explicitly comprehended; even textbooks of the so-called exact sciences verify this. Definitions are often missing, or are circular, and the study of a science consists, from a linguistic perspective, more of merely growing into a language than of a conscious learning of language usage on the basis of explicit definitions. The fact that there are philosophers of science who want to make a virtue of this historical defect, shall not be discussed further here.

Thus, for better controllability of spatial statements, it must be demanded that all relevant words be explicitly defined or terminologically determined in some other manner, even if the words in question are very common words of ordinary language or of a specialized science. The demand for an explicit reconstruction of the linguistic means and, with that, the distinctions which form their basis, can be focused even more by two additional conditions, especially for spatial words. This focusing is informed by the goal of expanding or of safeguarding the above-mentioned dual generality of knowledge. Spatial distinctions are,

1. to be made in *states of affairs of the outer world*; and
2. *artificially produced* states of affairs are especially good for this.

Concerning the first focusing proposition, especially in regard to the fundamental spatial properties of the world, there is a rich philosophical and psychological tradition which has been lent a new currency nowadays by biological "epistemology." According to this tradition, specific human thought-structures or dispositions, our so-called "*Weltbildapparat*" (Konrad Lorenz), the intuitiveness of our thinking, or any kind of inner, spatial self-experience, all play an important role in cognition. It shall not be asserted here that it is impossible to make valid statements about this. But one can easily convince oneself that all *statements* about the qualities or structures of our spatial thinking, conceptualizing, etc. are, trivially, dependent upon spatial *words*; always, in fact, upon such words as are used to describe outer, spatial states of affairs. If one asks about the possibility of definitional access to such words, it is not clear how a start could be made in defining mental or psychological operations, while, in contrast, it is completely unproblematic to begin with states of affairs which can be shown in the world. Provided with words which are based on the outer world, one can then speak about mental states or events. Even today, there are philosophers who think that after Kant or after Hegel nothing more of significance has been discovered. These philosophers tend, in the tradition of Kant, to deduce the possibility of linguistic communication between people (e.g., about an outer, spatial state of affairs, for instance, that the table which is currently being shared is square) from conditions which lay in the human disposition to thought. The argument, somewhat simplified, states that no linguistic communication could be possible if the speakers involved had not previously achieved something, like organizing their sensory impressions in a suitable way. On the other hand, it is undisputable that these statements (to the extent that the spatial words which are in question here appear in them), for their part, cannot achieve any definitional determinant in this way. Rather, such "epistemological" theses can only be formulated if the spatial distinctions which occur in them are already known.

It is not a case of some sort of philosophical chicken-or-egg problem here (which came first, the spatial word or the spatial conception). Instead, our context here solely concerns the better reliability of knowledge through the improvement of the comprehensibility of its linguistic framework. Theories about which mental or psychological processes must take place first, or simultaneously, in order that speech or understanding can occur, do not generate any reliability for a terminology, but already require a reliable terminology. The sole deciding factor is whether explicit and clear agreements have been made or not.

It is obviously advantageous when these agreements are made regarding states of affairs which can be shown in the outer world. When, for instance, the spatial word "between," as familiar as it might be for us in ordinary language, is to be once again explicitly defined in order to avoid misunderstandings, one can do this especially well on the basis of, for example, three neighboring pearls on a pearl necklace. In contrast, one will have a bad time of it if one attempts to derive the spatial "between" from a temporal "between" of states of consciousness.

The quality of the comprehensibility of a terminology is heightened even more when the states of affairs on which linguistic distinctions are standardized are *artificially generated*. Because, aside from the free availability - one need not wait until one coincidentally finds a suitable example for position, for a magnitude ratio or for a spatial form - the artificial generation or production of such a state of affairs offers the advantage of quite clearly generating the special aspect upon which the definition depends. This can already be illustrated by the example of the pearl necklace for the word "between": if the pearls are merely lying on the table without being strung together, then one would additionally have to talk about whether the form of their arrangement plays a role for the word "between." For instance, when they lie in such a way that each of the three pearls touches the other two - should it also be said that each lies between the others? Or should "between" only be used for arrangements that are approximately straight? When this application is intended, one would, again, do well to generate a suitable state of affairs, hence, to lay the pearls in a straight groove, etc. The advantages of the introduction of spatial words on the basis of artificially generated states of affairs can be seen even better in talk about the forms of bodies - who could renounce this approach with "circle" or "cube," with "regular hexagon" or even with "straight"?

The examples just given are supposed to suggest that states of affairs which are manually produced are what is meant by artificially produced states of affairs. Here, the contrast between natural/artificial is to be understood in the sense of the Aristotelian distinction between *physis* and *techne*, in which *techne* (τεχνη, dative!) characterizes something which is produced through human actions, and τεχνη (nominative) the art or talent in such production. Existing "by nature," in contrast, is anything which does not have its source in human intervention, but in itself. (The Aristotelian examples extend from the case of a stone to the growth of a plant, up to the motion of an animal.)

To make the definitional start with artificially generated states of affairs, and not with those coincidentally found naturally, is a methodical trick which is practiced quite successfully, even in physics. One may not, of course, consult the "official philosophy" of physics regarding this, which talks a lot about experience and natural laws. Even the self conception of the physicists is not a reliable source. One should follow the good advice of Albert Einstein, who said that if one wants to know something about physics, one must not ask the physicists, but instead must observe precisely what they really do. In so doing, one will see that long before physicists can collect experiences about natural events, they acquire artificial conditions in the form of their instruments. There would, in other words, not be a modern physics (from Galileo to today) if instruments for observations, measurements and experiments were not produced "first." (This "first" will be discussed in the next section.)

A criticism often arises here from a type of philosophy which steadfastly refuses to see the difference between stocks of knowledge which are *linguistically expressed* (and therefore tied to claims of truth), and *non-linguistic conditions*. Furthermore, this type of philosophy gives no account for the truth of its own statements. Generating artificial properties of bodies "already requires" that these be spatial or three-dimensional, or that we humans can move our hands in a three-dimensional space. In fact, some really cunning philosophers[53] even believe to have found an argument in the fact that our familiar everyday skills would not exist with a two-dimensional body, because then the neural connections of our brain would not be as they actually are in the three-dimensional case.

In answer, it must be said that, indeed, something comes before the explicit definitional standardization of words, which is, according to the proposal of manufacture made above, the state of affairs which serves as the basis for definition based on examples. It must be admitted that we enlist (non-linguistic) human skills in this. Something must truly be done, i.e., real bodies must be grasped, moved, sawed apart or otherwise worked. It would, however, be nonsense to assert that this "requires" that space be three-dimensional; the expression "three-dimensional," which we want to first gain in a controllable manner, already occurs in this statement. At least it would be a nonsensical circular definition, if one

53 W. Büchel (1965), pp. 151-156

considered the terms which one wished to define as already defined because of such "requirements."

However, the error actually lies even deeper than a circular definition: the objection above ignores the difference between a linguistically explicit *presupposition,* that is, one which is asserted and must, upon inquiry, be substantiated, and a non-linguistic *condition* (which is perhaps in retrospect discernible). The fact that the person acting is not stunned at the time, is also a "condition" for the manual working of bodies! But it is not at all true that this or other "conditions" must be named in valid statements in order for action to take place - which, it might be added, is quite fortunate. After all, what kind of world would it be, if people had to wait for philosophers to prove the possibility of their actions in advance before they could act?

The fact that one can move a body, or form one from clay, is numbered among the basic abilities of human beings, which we purely and simply lay claim to as a matter of fact, like anyone who is involved with a human practice, or like any philosopher who wishes to assert something about it. In this regard, *ab esse illatur posse*, possibility follows from reality, is the scholastic rule. The fact that we *can* act (and, in this, produce spatial states of affairs) follows from the actual undertaking, not vice versa. Of course, no objection is raised to the empirical research of such basic abilities or to analyses after the fact (e.g., in psychology), as long as a specifiable cognitive goal is pursued.

THE GENERATION OF SPATIAL FORMS UPON BODIES

"Spatial" has been established as a "meta-linguistic" word, i.e., a word which is itself to be applied to other linguistic objects. A short consideration of a few examples from ordinary language shows that we use spatial specifications for widely varied realms of objects. We localize bodies (billiard balls, for example) as well as events (e.g., storms, concerts, and wars). The transitions to spatial metaphors are fluid, e.g., the economy is in decline; the supervisor is behind his workers or places his interests above theirs; the president is above the chancellor; and the light is under a bushel. In view of this unassessable manifold, where should one begin?

One could be tempted to hope for aid from physics, since the physicists claim the responsibility of empirically researching space. However, they combine quite a variety of things in spatial statements:

First of all, one must disregard that an extensive use of spatial metaphors can be found in physics itself (thoroughly successful, incidentally, and applied with good reason); anywhere, in fact, where analytical geometry is applied. When, for instance, pitch, volume and tone color are used as the perimeters for the characterization of a musical tone in acoustics, or when hue, saturation and brightness are used to characterize a color, then one can speak, with good reason, of tone- or color-spaces. Such spatial representations of non-spatial states of affairs, however, shall not be considered here.

The fact remains that physicists make spatial statements in a stricter sense; they speak of distances and intervals, about the angle or form of curves, and, in the process, examine objects like light rays (in geometrical optics), fields (e.g., in gravitation theory or electrodynamics), temperature distribution, forces of tension in solid bodies, etc. Where does one begin, or where can one justifiably distinguish a starting point? The systematic proposal which is presented in this book starts with the *artificially produced form of bodies,* because this is a *methodical* start which leads to reliable spatial statements.

The assertion that this is a reasonable proposal shall now be further elucidated. It must be solely understood in light of the goal of finding reliable knowledge in spatial statements. Reliability alone is, however, neither an end in itself nor an isolated goal; otherwise, we would have to be on the look-out for the most logical truths possible, in that we, for instance, define "dimension" in such a way that a desired state of affairs is logically derived from it. (It is easy to demonstrate how unsatisfactory this method is: with the definition "dimension is the spatial property which is present exactly three times in bodies, events and empty rooms," the reliability of three-dimensionality would already be unshakeably derived.) The layman as well as the scientist has a certain pre-understanding of this problem, part of which is that the outer, real world and, along with it, *our experience of this world*, must somehow be taken into consideration. With that said, the problem which is now to be solved can be stated: How do we begin, in order to make our spatial experiences of the world reliable?

What it is that makes wholly personal and private experiences reliable for the individual (i.e., is distinguished by the fact that one actually relies upon it in one's daily practices) is a question which everyone must answer for oneself (and parents should consider this in the interests of their children). As explained above, we are talking here about experiences with the claim to being general, public and linguistically comprehensible. A "public" must be produced for experiences, i.e., there must be processes of linguistic agreement in which as many people as possible - in principle, all people (which is highly problematic) - can be included. ("In principle" is a limitation here, not an intensification: whoever has no interest in being included in an agreement process, or whoever is too stupid, needs not be further considered; this does not lead to the breakdown of the claim of the comprehensibility of an experience for "all" people. We want to leave it open here, as to where the claim actually does break down. We will only make a proposal here as to how an improvement of agreement possibilities can be reached through an explicitly described process. The highly important cases in which attempts at such processes fail will not be considered here, as they have nothing to do with the elementary spatial experiences that are discussed in this book.)

All speech about the commonly experienced, outer world begins unavoidably with sensory perceptions (and not, of course, with the linguistic sentence, "It begins with sensory perceptions"!) In other words, we speak about what we see, hear, etc. All criticism concerning the reliability of sensory perception, and all attempts to replace this through "objective procedures" with instruments - for instance, one mistrusts the statement that this rose is red, and therefore measures the color spectrum which emanates from it in sunlight - misjudges the fact that we would, in general, have absolutely no possibility to talk about the world if we had not gained the possibility within the cultural history of language, or simply if we did not find the possibility today, of *agreeing* about our perceptions. We would be even less in a position to make any *theories of perception* or about the physical properties of perceived objects, if this possibility of agreement were not available to us. Hence, at the (methodical) beginning, the preparation of words about "simple" perceived spatial properties of bodies and perhaps other objects (e.g., a rainbow).

The reference to "simple" perceptions, i.e., those without technical aids, has a disadvantage which can not be overcome. In the case of a difference of opinion, the means of testing and agreement are very quickly exhausted. What, indeed, should be done when two people in the same situation are of different opinions concerning whether two objects are the same color, or if two tones are equally loud? The situation becomes even more problematic when the dispute is a matter of whether two people have received equal amounts of water or gold. Such problems of the just division of goods were probably the beginning of human efforts to make rational decisions possible in place of authoritarian or jungle-law decisions. Instruments were invented to serve justice. (Symmetrical scales were already a symbol of justice by 2,500 B.C., aside from their useful application in the comparison of weights.)

One need not immediately think of measurement, nor even of modern science, when one thinks of the application of instruments which supports the (still indispensable) sensory perception. Because measurement was not developed for science, but for practical purposes, long before the historical start of the scientific art of measurement. And, (methodically) before the practical art of measurement, came other capabilities or possibilities of instrumentally-supported controls on states of affairs about which no agreement, or at least no easy agreement, can be achieved through sensory perception alone. Whether or not, for instance, a star lies between two other stars with the extension of the straight lines, is then no longer decided with the fist, or with a claim to special competence or authority, but by using a sighting ruler: if this is straight then the question can be decided through a process which is independent of star-gazing.

This small digression in epistemology and the theory of science is supposed to show that we

1. unavoidably begin with sensory perception,
2. but can then expand our control and agreement possibilities by adding instrumentally-supported experience.

Whether or not agreement can be reached about the instruments themselves, i.e., about their necessary properties in concrete application instances, obviously plays an important role. Of course, a sighting ruler does not help in star-gazing when there is no agreement about what "straight" means or about how one tests the straightness of a ruler. (Expedient to this end, one might consult a second ruler and test whether both rulers fit each other around their edges.)

Instruments are, aside from whatever else they might be, artificially worked bodies. It is therefore human actions which bring about (and in this sense, determine) the respectively necessary and demanded properties of the instruments. Instruments or apparatus are produced and oriented for a particular purpose of application. The possibilities for agreement in reference to sensibly perceivable spatial properties are thus improved through the use of instruments, which were generated solely for this purpose. Entire museums bear witness to the cultural history of the art of the manufacture and use of rulers and protractors, of clocks and telescopes, of field measuring instruments and astrolabes, of scales and thermometers, etc.

The question about which object we should begin our definition of spatial properties with, in the face of our manifold possibilities in ordinary language and the manifold physical realms of spatial speech, has therefore been answered. Whatever else we know, and must know, in order to attain reliable spatial statements, to the extent that we only wish to enlist appropriate instruments for the improvement of control and agreement possibilities, we must *begin with the technical mastery of bodily forms*. Otherwise, we would have no art of instrument construction nor, in turn, control of the desired properties of the instruments.

The historical fact that modern physics long ago passed the stage at which all phenomena are supposed to be traceable to the spatial properties of bodies (or any properties of bodies), is not an objection to this beginning with artificial, spatial bodily properties; since instruments must be worked with whenever physics (whether microphysics or astrophysics) seriously wants to be a science of experience or laboratory. There would be no physics if there were no successful use of instruments, and therefore no technical mastery over the spatial properties of bodies. However, instruments are also, of necessity, always spatially formed bodies.

POIESIS AND REENACTABILITY: THE PRINCIPLE OF METHODICAL ORDER

Derived from the Greek word "*poiesis*," meaning manufacture, those actions in which objects are produced manually are called "*poietic*." With this word, Aristotle wanted to direct attention to a difference between this and practical actions. Of course, actions like going for a walk, holding a speech, electing a representative, or catching a thief, are also actions in which people pursue particular purposes. But poietic actions occupy a special position. They lead to products with a lasting autonomous existence. It is of particular importance, concerning the spatial properties of these products, that the permanent properties are determined by the purposes and intentions of their inventors and builders, and in the second phase, their users. The natural properties of such instruments are either not at all determined by such purposes, or are only indirectly influenced. Of course, such natural properties are *also* always present, e.g., a ruler has a weight, and also expands in heat; these properties are not intended by the user of the instrument, but must simply be accepted.

The enormous effect that modern natural science has had on the understanding of the world of our time has, above all, directed attention to aspects of natural law and our dependence on experience for our knowledge. But like every successful beginning in the history of human efforts toward cognition, natural science has also led to wild exaggerations. Of special note here, is the attempt to misjudge the *possibility of natural science*, more precisely, its character as cultural achievement, and to instead try to give an explanation for this from natural science itself.

If it is admitted that natural science would not be possible without the successful use of instruments, then the next question is, what, in turn, makes the successful use of instruments possible? According to this doctrine, this is, again, a question for natural science. What this concretely means is that one "explains" the functioning or the properties of the instruments used by physicists in terms of natural law. One might say then that the natural law of pendulum oscillation is realized or "works" in a pendulum clock. This is the (empirical or naturalistic) rival view to the view which is developed here, according to which the properties of instruments - the significant ones for which the instruments are built and used - are only comprehensible in terms of human purposes. Which of the two views should be preferred, since they are not compatible and cannot both be valid?

In any case, the view of natural science (or more precisely an expansion of natural science raised to the level of philosophy or epistemology), runs into the following dilemma: The *appropriate instrument* for a physicist or a layman can only be decided in the context of application. It is not enough, for example, that an instrument looks like a common clock, or is simply called "clock," if its hands do not move. It can be immediately specified by any competent user (his competence consists of precisely this) for any given instrument, just which properties or functions must not be missing in order for the instrument to still be able to fulfill its purpose at all. Now all kinds of defects can occur in instruments. Defects are recognized as such, precisely because the instrument fails in its purpose. In other words, defects in instruments are always the failure of the purpose of human actions. On the other hand, and this is the dilemma of a natural law explanation of the properties of instruments, they are not a refutation or failure of natural laws. Defects in instruments are also "explainable in terms of natural law." When the weight of the pendulum clock has run down, and the clock comes to a standstill because of it, of course no natural law is refuted; but, indeed, the stopped clock fails the human purpose for the use of a clock. The conclusion which can be drawn from this is that it is *not natural law* properties, but *properties which are defined through human purposes*, which are realized in poietically generated objects, in instruments.

That is why, in spite of natural law descriptions, spatial properties of bodies must be discussed under the aspect of their dependence on action, that is, in the scheme of means-and-ends relationships. Purposes often lead to restrictions for poietic actions which must be especially taken into account: not every sequence of single actions in a chain of actions is successful. An example which goes back to Hugo Dingler has been chosen as an illustration. This will also be used in establishing a principle which will be applied in the theory of three-dimensionality which follows.

Whoever wants to produce a painted wooden figure, and has thus made this his purpose, would do well to first carve the figure, and then to paint. This sequence is not prescribed by natural law. Nothing prevents one from carrying out this action in the reversed order. Since we are interested in reliable knowledge about space and, with that, about reenactable spatial statements, it is decisive that the *poietic* actions of carving or painting are *spoken* of. On the other hand, nothing prevents one from linguistically comprehending the sequence of these actions differently, that is, to describe or (with intentions of definition) to regulate it differently. But one knows, of course, that descriptions which deviate from the single successful sequence are false, and that deviating rules fail. Therefore, let us establish a "*principle of methodical order*" which *forbids* certain *linguistic procedures*: it is forbidden, to describe or regulate *poietic* actions in a sequence different from the one which must be carried out if the action is to reach the intended goal. The principle of methodical order regulates linguistic steps, or, more precisely, it forbids certain sequences. It therefore also forbids, as one can easily illustrate for oneself, the circular definition.

An example might illustrate that this principle is not trivial, for instance, in that it is always followed anyway: when the second law of motion of Newtonian mechanics (in the usual textbook abbreviation "Energy is mass times acceleration") is read as a definition for "energy," according to which the energy affecting a test body is determined by the measurement of the acceleration of the body with a (measured) mass, and when, on the other hand, this acceleration measurement can only be carried out within a certain reference system in which force-free bodies in a state of rest or of equilibrium pause, thus are not accelerated, (the first Newtonian law of motion), then the act of measurement could not really be carried out. One would already have to be able to measure forces, in order to be able to produce an appropriate reference system, and vice versa, would have to have an appropriate reference system in order to be able to measure forces. Here is a circular definition which is forbidden precisely because it does not correspond to a successful sequence of poietic actions.

Principles like that of the methodical order prescribe something, they are normative. In areas of life in which no authority with law-making jurisdiction is established - and philosophical critique is the realm par excellence in which the establishment of such authorities is forbidden - a principle or a norm must be *justified*. The justification here turns out to be rather harmless, and does not lead to profound foundational problems. Here a relative justification should be sufficient: violations against the principle of methodical order lead to false assertions in descriptive speech and to misleading instructions in regulative speech. So whoever does not wish to accept these consequences, or to bring them about himself must adhere to the principle. But the compulsory nature of the principle of methodical order cannot be impressed on those who do not care about the factual reenactability of asserted actions and, with that about the truth of his statements in regard to these. Nor can it be brought home to someone who, for example, does not shun giving "operational" definitions of physical concepts in physical textbooks in such a way that one cannot follow them in reality.[54]

In summary, we have now set up and established a catalogue of measures which should guarantee the reliability of our knowledge about space, to the extent that it is formulated in the following theory:

It is a matter of *linguistically* formulated knowledge maintained in an *expressly determined terminology*. This is, beyond the explicitness of the definitions, a *reenactable* knowledge since all spatial distinctions are *methodically ordered*, i.e., they are determined through factually executed poietic actions for the production of precisely those spatial states of affairs which are then spoken of. And with that, despite possible further experiences of physics, for instance, a methodical start with the working of bodies has been distinguished.

In conclusion, the question could be raised about why one should start with the form of bodies (and not the magnitude or the position, as the successful application of analytical geometry would suggest). This question was already warded off in the discussion of Galileo`s contribution to the theory of three-dimensionality (Cf. Part I, Chap. 1). Now after the formulation and establishment of the principle of methodical order, the reason for this rejection can be given more clearly and briefly: the forms of bodies can be technically and terminologically mastered without the aid of measurements and quantitative details (as can be read in the relevant literature to the foundation of geometry, and as will be verified in the next section), while this is not true for definitions of magnitude. In other words, there are complete, non-circular definitions of forms and then, with their help, of magnitudes (like "length" or "equally long"), on the basis of procedures which can be actually carried out. In contrast, the reversed method, of first defining magnitudes and then characterizing forms with magnitudes (e.g., the straight line as the shortest connection) cannot really be carried out. The fact that the latter method is nevertheless often favored is based on a false interpretation of the history of physics.

---

54 As when a "conceptual operative definition" is suggested by P. Mittelstaedt (*Klassische Mechanik*, Mannheim, 1970) without a corresponding procedure of realization - e.g., for inertial systems. This leads to a complete misunderstanding of actual operational approaches. (Cf. P. Mittelstaedt: *Der Zeitbegriff in der Physik*, Mannheim/Wien/Zürich, 1976, p. 145.)

CHAPTER EIGHT

The Construction of the Terminology

BODIES, CUTS, SPATIAL FORMS

Now a terminology shall be "reconstructed" step-by-step, according to the principle of methodical order, with the goal of providing reliable knowledge about the three-dimensionality of space. The word "reconstruct" is supposed to suggest that it is not a matter of inventing an artificial language here, but merely a matter of explicitly reenacting distinctions which are already known. In so doing, ordinary language must be taken into consideration, as well as the usual distinctions of the relevant branches of science. One must certainly pay heed, for example, to logical or methodical incompatibilities in the familiar ways of speaking. Thus, the greater exactitude of an explicit regulation of language often requires a deviation from accustomed language usage.

Until now, it has been argued that the reconstruction of the terminology has to begin with artificially produced properties of bodies. However, a terminological specification for the word "body" has not yet been considered. This must now be remedied.

An extensive debate could immediately begin - regarding the philosophical tradition, again - as to what the relevant aspects of corporeality are, like imperviousness, gravity, or inertia (in the physical sense), materiality or some other property. But, within our context one must not concern oneself with these questions, which are often raised in a conceptual fog which only favors the violation of the principle of methodical order. After all, it is a matter of the production of instruments through the working of bodies: instruments, that is, which then make the gaining of scientific experience in the form of spatial statements possible. The constructor and builder, as well as the user, of such instruments, knows what he must think of when bodies are required for the manual working and production of tools and instruments. It is neither a matter of individual atoms, nor of entire planets, neither of objects like clouds or gases, nor of fields or light rays. The pieces of material which are to be used must not be too small or too large, neither too hard nor too soft. Hence, "*body*" should be understood in an everyday sense, which assumes that one knows how bodies must be worked and moved.

An elementary possibility of artificially producing spatial properties upon bodies is to cut something up (for example, with a knife or a saw). The word "*cutting*" can be defined with examples of the action of cutting. When we learned words for actions as children, we learned them in the context in which these actions were carried out. The teaching and learning of action words is reasonably conveyed in connection with the teaching and learning of the corresponding action. Every master craftsman follows this method in instruction. In any case, every competent speaker is then also competent in judging whether this action can be carried out at all, or correctly, in each concrete case.

The result of the act of cutting through a body shall be called a "*cut*." In order to avoid having to take a number of special cases into account, it is assumed that the body which is cut, as well as the cutting tools, are not extravagant, but are as *simple* as possible. Thus, one does not begin with a rounded knife, such as one might use for cutting out the core of an apple, or with a body in the form of a pretzel or a doughnut (ring). Rather, a cut should divide a body into exactly two parts. (This condition will be explicated more precisely later.)

Something new originates on the parts through cutting, namely, "surfaces." Following the consideration that the reconstruction of the terminology should start with artificially produced spatial properties, the normal language usage, according to which every body bears a surface even before it is cut, is not adopted here. Rather, *surfaces originate through a cut* per terminological stipulation. (Normal speech about the outer surface of a body, like that of a pebble, is only later derived from this definition, understood as a type of "as-if speech," namely "as-if" the outer surface had originated as a cut.)

Both parts which originate as the result of a cut have a surface. We will assume concerning cuts and cutting tools, that the parts which originate *fit together* at the cut surfaces, or briefly, *fit*. (One could think of an apple that is cut through by a sharp knife, so that the two halves can again be put together, and would stick together with adhesion.)

Hence, it is *assured by definition* that the cutting of a body is a suitable action for producing a pair of matching surfaces. In our everyday experience there are many examples of matching surfaces which did not originate from cuts in the sense of the previous definitions. Screws and their matching nuts are not produced by being cut out of one piece of metal in one and the same operation. Some materials are not at all cut, but are poured, ground, etc. When, for instance, an achromat (a lens system which avoids color flaws in the image) of individual lenses are fit together for the objective lens of a camera, this is the result of a costly grinding process, whose goal is, among other things, the matching of separately produced parts. Taking this into account, *surfaces* shall now be used to mean, in an expanded language usage, *all artificially produced* surfaces, even if the procedure normally is not called "cutting." And the *fitting* together of surfaces shall also not remain restricted to the origin from cuts.

After this expansion of the manual procedures which are admissible for the terminology, we can now return to the generation of new spatial properties. If a part of a simple body, which originated through a cut, is cut in such a way that again two parts come into being (of which both now have part of the first surface) then a new spatial state of affairs has emerged: each of the two new parts has an "*edge*."

If a part which has an edge is, once again, cut into two parts such that each of the new parts has a piece of the edge, then a "*corner*" comes into being on both parts of this (third) cut. Surfaces, edges and corners are examples for spatial "*forms*."

Looking back at the first part of our terminology, we now have the following:

1. the word *body*, determined by examples within a context of means and ends;
2. the words *to cut*, *cut*, *part*, and *to fit*, determined within contexts of production;
3. the words *surface*, *edge*, and *corner*, determined through definition, as well as their collective term, the word *form*.
4. Additionally, "cutting" and "cut" are to be expanded to include surfaces, edges and corners, i.e., bodies can be cut, and cuts can be determined, in such a way that talk of cuts through (or of) surfaces, edges and corners is understandable.

If not already asserted on the basis of sophisticated considerations within some theoretical contexts, then certainly no one who has learned the cutting of bodies and the corresponding language would doubt that one can also cut through bodies at the corners. Anyone can perform this himself after having read the preceding sentences. But the question can arise regarding the reconstruction of the vocabulary as to why the list of words breaks off at the word "corner." After all, one could name the form which comes into being from a cut through a corner, "tip," for instance. One could also search for words from ordinary language to name the fifth and further cuts, or one could use indexes for counting cuts, and could speak of "n-corners." But, in fact, as known in ordinary language, corners are spoken of regardless of the number of surfaces or edges which come together in them.

In other words, there is a decision which must be made here in the reconstruction of the terminology, which has to do with the adequacy of ordinary language, as well as that of the language of geometry. A close connection with the problem of three-dimensionality is obviously exhibited here. If exactly three - and only three - spatial forms can be generated through cuts or (synonymous to this) through division, and if these are understood in connection with the ancient way of thinking as boundaries (corners bound edges, edges bound surfaces, and surfaces bound bodies), then the three-dimensionality of bodies in the sense of a triple divisibility is already established.

But one must bear in mind the *arbitrariness* and free stipulations which are responsible for this understanding of three-dimensionality: this fully arbitrary decision to bind the decisive spatial terms "surface," "edge," and "corner" to the act of cutting, but to prohibit any new words from the fourth cut on. Such arbitrariness is not justified by the fact that it is of venerable age. The ancient geometricians had already taken this path by characterizing the regular polyhedra (the "Platonic bodies") by the indistinguishability of the corners, edges and surfaces, without naming reasons as to why this enumeration should be complete (Cf. the discussion in Part I, Chap. 1). The completeness of the enumeration was, of course, adopted from ordinary language and was considered previous knowledge of the geometry pupil, not worthy of discussion. For instance, the first definition of Book XI of Euclid's *Elements*, in connection with the ordinary language of the time, also only repeated the definition that a body is that which has length, breadth and height. The requirement of indistinguishability of all corners of a regular polyhedron, or more precisely, the equal treatment of the polyhedra corners of all five Platonic bodies inherent in this requirement, is not self-evident. Because the corner is formed by three surfaces, or three edges, in only three of these polyhedra, while in the octahedron each corner is formed by four edges or surfaces, and in the icosahedron five edges or surfaces form the corners - which of course must be taken into account in Euclid's proof.

As far as we can tell today, the ordinary language in the time of Plato or Euclid had already reached the point where, for the geometricians, there were no proper words for corners which had been multiply cut. If one questions the source of the layman's intuitive conviction concerning the certainty of the three-dimensionality of space, one might sense a possible answer here - to the extent that one acknowledges the connection between the triad of these forms and three-dimensionality. However, the goal of providing reliable knowledge about space must bring about an insight in place of the uncritical adoption of ordinary language. Beginning spatial terminology with cuts through bodies does *not* provide an argument for three-dimensionality; this is, rather, a fully arbitrary decision.

However, it is easy to see why the triad of these forms has successfully asserted itself, in spite of this arbitrariness. Since antiquity, one has understood a point to be an indivisible form. (Euclid's definition: A point is that which has no parts.) One applied speech about dividing, or cutting, in an unclarified manner, to both the realm of real bodies and to that of abstract geometrical forms. This perspective forced one to dispense with further terminological considerations of the division of corners. While surfaces on real bodies correspond to surfaces on ideal geometrical forms (so much so that neither the Greek nor English languages have two different words for them), and edges on real bodies represent geometrical lines, real corners stand for ideal points. But because one calls points "points" regardless of whether two lines, or more than two lines, intersect in them, one also applies the word "corner" regardless of the number of edges or surfaces which come together in it.

We do not want to go the way of simply following the tradition here without any reason; nor do we wish to contradict it without reason. Rather, one must first see that a decision in favor of three-dimensionality at this point of the terminological reconstruction would be completely arbitrary, resulting from a mere definitional stipulation. Second, it must be questioned whether the well-founded construction of terminology for the spatial forms of real bodies with the goal of reliable knowledge has already reached its end. This question can be answered with a clear negation. Our current vocabulary is still too imprecise and is not encompassing enough.

HALVING BODIES

Until now we have referred to the familiar everyday practice of cutting bodies and, as a restriction, have only required that the bodies, as well as the cutting tools, not be "too extravagant," e.g., neither should be ring-shaped. Every cut should divide a body into exactly two parts. Of course, the straight knife and the plane cut play an especially important role in real practice, so we will now turn to the definitions of these forms, which are also distinctive in geometry.

If one proceeds in the sequence of definitional steps as found in Euclid (and in accordance with Aristotle's theory of definition), then one must be led to the notion of first defining "straight" as the predicator for knife, and then for the direction of a cut, in order to then derive the plane as the surface of the cut, which originates from a straight cut made by a straight knife. But one immediately sees that, in this way, proceeding from previously determined words, one only exchanges difficulties. If the knifeblade is considered as an edge, one knows from simple common knowledge that straight edges are formed from plane surfaces. Hence, it is incomprehensible that, "plane" should be defined by straight. And above all, what would it mean to make a straight cut if one already had the straight knife? It could hardly mean to make the cut along a straight edge which one "projects into" the inside of the body to be cut. There would have to be at least two parallels, imagined edges or lines to accomplish this, i.e., one would need a further definition of form. The consequences of the preliminary considerations concerning which measures make spatial knowledge reliable become apparent. The vocabulary for spatial statements shall be reconstructed, step-by-step, from artificially produced distinctions which are bound to the principle of methodical order. These maxims now lead to a *deviation from the Euclidean-Aristotelian sequence* and to begin by distinguishing certain surfaces, and not to start with edges, or even corners.

A surface and, correspondingly, a cutting and a cut, are called "*plane*" when two surfaces which fit it also fit each other. Of course, this definition proceeds from the assumption that we now have three parts to consider, with a cut surface each. When these fit each other pairwise, then each of them is called "plane," which results logically from the definition.

However, an important addition must immediately be made here, so that surfaces like regular corrugated tin or saddle surfaces are excluded, since these, too, fulfill the condition of the pairwise fitting of three samples. One can demand, for instance, that the fit be maintained even with any given turn or shift of the parts relative to each other and that, in this process, any overlapping parts are to be disregarded.

Incidentally, there is, in this, an impressive superiority of ordinary language means over mathematical-logical explication. One can think of, say a raw potato cut through very precisely to produce plane cuts. According to the definition, we say that the two parts originating from this cut fit each other. We can lay them together in such a way that the boundaries of the surfaces come together. But it is also immediately clear to us that the fit is maintained during a shift, although there are, perforce, parts on both surfaces which do not touch the opposite surface, but overlap it. Feeling quite certain, we distinguish this case from one in which a polyhedron, like an icosahedron, is laid on a plane surface, and then only a triangle fits. In this case, one would not say that the triangles bordering the first were merely overlapping parts, so that there were a total of four connected triangles which fit the plane. On the other hand, it is very awkward to define, in a logically and formally correct manner, what it meant by saying that the fit is maintained during a shift, and that one is to disregard any overlapping parts, since this definition cannot make use of the word "plane." This little digression about the advantages of ordinary language over an exact terminology should provide reasons for further maxims to follow which take into account the ensuing terminological construction and the theory which is developed from it.

Explication or explicitness in the reconstruction of the linguistic means is not considered here to be an end in itself. So it is rejected here that a theory or a terminology is better the more detailed its definitions are. Instead, detail will only be gone into to the extent that it is useful for the goal of the approach and of the book. In other words, to the extent that talk of dimensions in an everyday and in a scientific sense can be clarified, and the triad of ordinary space can be proven. The definition-theoretical practice of the mathematician - as in a modern textbook on topology - shall not serve as a model here. The low formal exactitude which must then be tolerated, is weighted against a significant gain in epistemological exactitude, which is formed by the embedding of the linguistic means in the technical context of the production of spatial forms on bodies.

It is perhaps conspicuous that something new has been brought into the game here; namely, the relative shifting of two surfaces or bodies. This can only be justified later (Cf. p. 190). Shifting plays a role in the generation of certain spatial forms and the ensuing assertions of unequivocality, as well as, finally, the definition of the central term "dimension." Whoever suspects a violation of the principle of methodical order in this anticipation of free shifting can, however, be reassured. The *definition of "plane"* for cuts and surfaces is *not* dependent on shifting, as will now be shown in a tightening of the definition for fitting.

"Fitting" was determined for surfaces from one and the same cut. It follows that the fitting surfaces are even "*deckungsgleich*" (they cover each other exactly), i.e., no surface or part overlaps the others. "*Deckungsgleich*" must not be confused with the geometrical term "congruent," since the surfaces which cover each other are surfaces on bodies, and therefore (can) have different spatial orientations. A restriction to plane surfaces has not been made until now. Therefore, this concept of fitting contains the silent assumption that the parts which are divided by a cut, can be reassembled again in just the same way as they originally were connected.

Since we now want to dispense with the special requirement that the cut bodies not be formed in such a way that there is more than one possibility of laying them together (e.g., when a circular or square cut surface results), we will now demand that marks be made on the boundary of both cut surfaces which must come together when the parts are laid together. If this condition is fulfilled, then the *fit* is called *oriented*. Wherever something else is not expressly specified, it is always the oriented fit which is meant.

The following is then required in the definition of "plane" for the third body with the fitting surface: for both fitting parts which originated from the cut of a body, one chooses, as a third, a "larger" body, i.e., one which completely overlaps the fitting surface of both other parts. Then both coinciding surfaces can be laid on the larger one (one after the other), and can be traced around with a colored line. The mark which was demanded above can likewise be drawn onto this colored line, and this must, again, be properly taken into account when laying the parts together. Of course, one must not expect that both coinciding surfaces be reproduced "equally" in this way, i.e., that the same lines result on the third body. Rather, depending on the form of the original body, the opposite could be the case. If a geometrical terminology were already available here, one could formulate it by saying that both reproductions are mirror images. It is, of course, decisive for the definition of "plane" that both cut surfaces drawn on the third body cover as much common area as possible, otherwise the definition of "plane" is violated.

This complication of the definition is obviously unavoidable. Without the orientation of fit, a large class of various saddle surfaces, as well as regular corrugated tin, would also be examples which fulfill the definition of "plane," and would generate a unacceptable collision with familiar language usage.

A spatial form is defined with *plane cuts*, or *plane surfaces*, for which a striking *replaceability* is given. In reference to the property of the mutual fit, any two of three plane surfaces can replace each other, in the following sense: one would like to produce a triplet *a*, *b*, *c* of plane surfaces. One brings *a* and *b* in contact. They fit each other as required. Then one removes *b* and puts *c* in its place. From the pair which originates from this, one now removes *a* and puts *b* in its place. One has then *"replaced" a* with *b*, as it is now defined (Cf. Fig. 5).

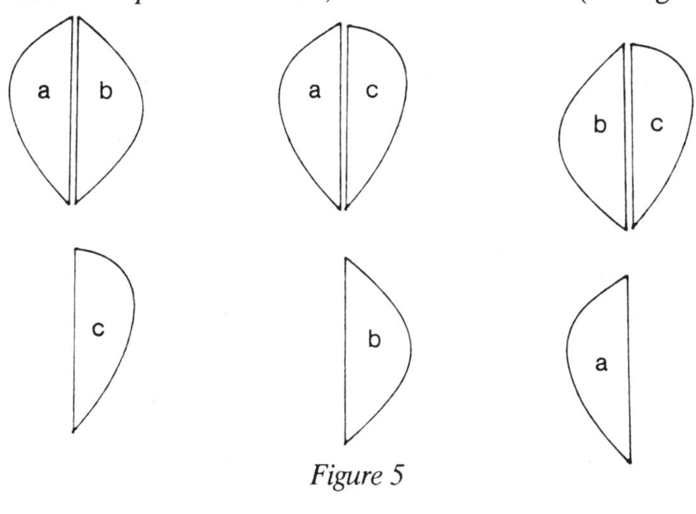

Figure 5

The criterion of the fit enters into this type of replaceability, i.e., one can say that two surfaces are *replaceable in reference to fit*. Now, it corresponds to normal language usage to call objects which are replaceable by each other in reference to a particular criterion, "*equal*." Equality is not bound to quantitative criteria. Two houses, two tones, two meals, two theater performances could also be called equal, and normally the kind of equality which is meant follows from the context, i.e., which criteria are applied. Hence, talk of equality does not at all require indistinguishability in all known or conceivable respects. (In the case of equality in every possible respect, that is, in every respect which can be formulated with predicators, one speaks rather of identity; one simply has two names for one and the same thing.)

For the sake of abbreviation, we shall now introduce talk about the *halving* of bodies. If a body is cut through in such a way that the generated parts are equal, then this body is called *halved*. Since the normal practice of halving almost always has a quantitative reference (when halving bodies, one is, as a rule, interested in *equally large* parts, i.e., equality in reference to magnitude), it must be expressly stated here that plane cuts for the halving of bodies do not concern volume at all. Per definition, halving cuts generate equality in reference to the fit or, more generally, to the spatial form. In competition with ordinary language, this form-theoretical language usage of halving might seem disconcerting at first; on the other hand, it is normal to say that the plane halves space, in which not the equality of magnitude, but the indistinguishability of the half-spaces in the form-theoretical respect is meant.

It is easy to see that "halving" is, again, a word for a human action, and that "halved" is an adjective for the result of an action.

What was indicated provisionally above, as reference was made to "simple" bodies and knives which are "not too extravagant," can now be explicitly defined. A *body* is called "*simple*," when:

1. Every possible, plane cut through the body results in exactly two parts, and when
2. every possible, plane cut through any given number of parts of the body resulting from plane cuts, also only results in two parts.

As an illustration of this definition, but not as one of its components, it must be added that, according to definition, only bodies which are plane or convex at every place of their outer surface are considered "simple." A body must not exhibit a "nose" which could be divided by a plane cut in such a way that, besides the "nose" itself, another part would also be cut, so that *three* parts would originate from the cut. Nor may a body contain a "hole" or a "dent." If this were the case, a plane cut could generate a ring, which could be divided into two parts by a plane cut in such a way that at least one of these parts would fall into *three* parts with a further cut (Cf. Fig. 6).

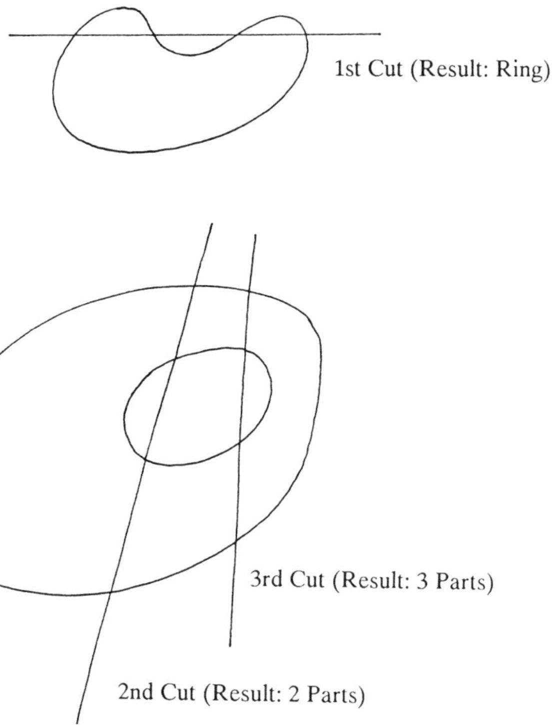

Figure 6

The process of halving can be applied to bodies which have already been halved, so that after the surface which has been distinguished as plane, other special edges and corners can also be more closely determined in definitions:

An edge is called "*right*" (or a "*right edge*"), when it results from the halving of a part which resulted from halving. That is, one proceeds with a body which is a part from the halving of a simple body, which thus has a plane surface. One then cuts this with a plane cut in such a way that 1. the existing surface is divided (and an edge is thus formed according to definition); and 2. the (new) parts are equal. Intuitively stated, this is a wedge whose two plane surfaces form a right angle. As one can easily illustrate, a replaceability analogous to that for plane surfaces also applies for these right wedges, with the addition, of course, that the edges which originated from the second cut coincide, or that the two touching wedges which fit one another must lie on a plane surface (which is the same thing).

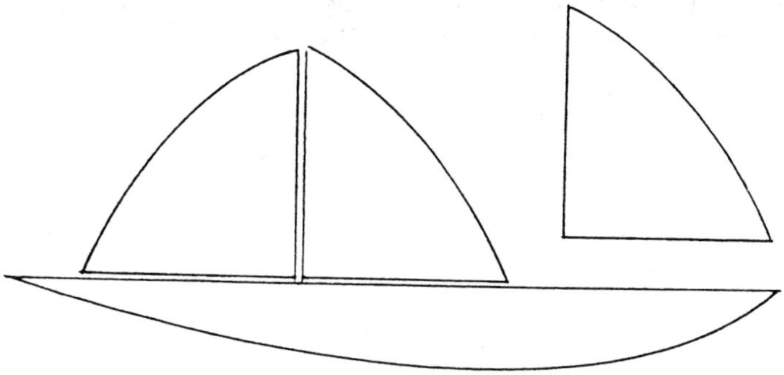

Figure 7

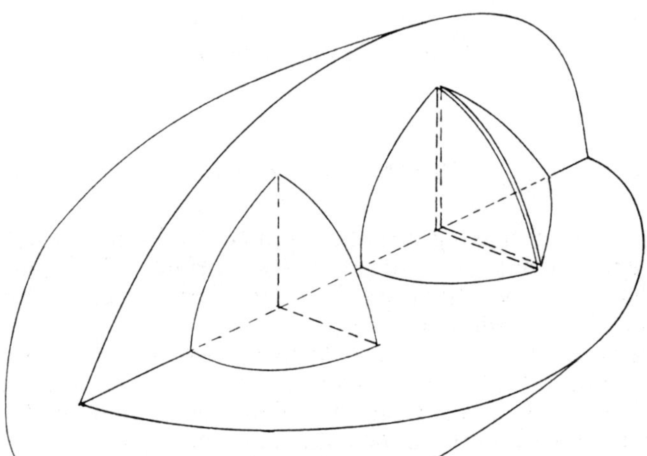

Figure 8

Again, one needs a third right edge on a third wedge, in order to determine the replaceability (and, with that, the equality). (Cf. Fig. 7)

A corresponding definition can now be given for the corner: a corner is called "right" (or a "right corner"), when it results from the halving of a part which resulted from a double halving cut. (Cf. Fig. 8)

It will be shown below that it must be terminologically distinguished here between the characterization of the respective form and the (part-)body which bears it. We will therefore name a body which bears a right edge, a (right) *wedge*, and analogously, a body which has a right corner will be called a (right) *vertex* (*Spitze*)[55] - following everyday language usage which distinguishes between sharp and pointed, applying the former to edges or wedges, and the latter to corners, cones, etc.

For the definition of the right corner one proceeds - in analogy to the definition of the edge - with a right wedge, and cuts it in a plane cut such that the resulting parts are equal, hence replaceable, which can then be defined or controlled with the help of a third right vertex.

In defining the right edge and the right corner, only *one* part was ever considered from the preceding cut. The other was laid away, as it were, and not worked upon further. In order to be able, in what follows, to examine the results which arise from cuts through cuts, this restriction shall be lifted and a terminological expansion will be gained in this way.

THE CONTINUED CUTTING OF BODIES

The cutting processes, or cutting tools, were intended to be such that the parts which originate from a cut can be fit together (oriented) again afterwards. We will assume that no essential technical problems arise when we cut bodies which have been reassembled anew in this way, as if they were undivided, and then proceed to ask about what it is that results from the new cuts. This corresponds to an expansion of our language usage about cuts, which no longer only leads to pairs of surfaces (as up to now), but possibly to four edges, or perhaps eight or more corners. We therefore use the following definition:

Cuts (and accordingly, cutting or dividing) are called "*continued*," when all parts from possible preceding cuts are fit back together

---

55 Translator's note: The German word "*Spitze*" is normally translated with the words "peak" or "point," as in the peak of a mountain, or the point of a pencil. "Peak" is simply wrong in this context, and "point" would lead to precisely the type of terminological confusion which Dr. Janich is attempting to dispel.

before further cuts are made (through these bodies which are fit together).

This definition has been made so that the first cut through a body is likewise counted to the continued cuts, and must not always be excluded, or discussed as a special case. Where there is no preceding cut, then the condition that parts from earlier cuts be fit back together is, of course, empty.

The definition, however, in no way demands that later cuts meet, i.e., cut through, forms from earlier cuts. A classic case where this is not fulfilled, is the cutting of bread into slices. All of these cuts (plane and parallel, in the case of those generated by machine), are by definition "continued."

If it is additionally called for that every continued cut meets the forms from all the earlier cuts, this is then called *"strictly continued."* (Nothing further will be determined or demanded for the general case of cuts, i.e., when a certain form of cut is not specified, about the particular kind of "meeting" of cuts, whether in a part surface, an edge, or a corner.)

Both terms, "continued" and "strictly continued" are also applicable to halving cuts. Accordingly, a *halving cut* is called *continued*, when parts from possible preceding halving cuts are reassembled before the new halving cut is made. And then *halving cuts* which meet all preceding surfaces (from halving cuts) are called "strictly continued."

With that, we now have all the necessary terminological means available for formulating and proving the decisive theorem for a theory of the dimensions of empirical space.

THE BASIC THEOREM OF DIMENSION THEORY

If we review the terms which have been brought forward until now, talk of surfaces, edges and corners has been made possible (without distinguishing special forms of cuts, for the time being), without having won an argument for three-dimensionality, or even a meaning of the word "dimension" from this. Using these terms, plane cuts were defined, and with their help, in turn, the right edge and the right corner. This made it possible to speak of form equal parts, in the sense of their mutual replaceability with the help of a third part, as well as the classification of cuts as halving cuts. The multiple cutting up of the same body, which is reassembled after every cut, and a possible overlapping of cuts resulting from this, with the help of the terms "continued" and "strictly continued," should open up a way to draw logical conclusions about other parts from the properties of one. This possibility will now be pursued only in regard to the most strongly focused use of this terminological means, namely, the "strictly continued halving cuts." In this way, one has come to the linguistic possibility of cutting a body up into four right wedges, or eight right vertexes. There are at least two things which are striking about this.

1. This series of definitions has made assertions possible which obviously demand proof. For example, the definition of the right corner is formulated in such a way that the three corners named in the definition are equal in the sense of their pairwise replaceability. In the triple strictly continued halving cut, however, we have arrived at eight right vertexes or corners. Are these also equal in the sense of the definition? In other words, may one assume that each of these eight vertexes is replaceable by every other one? One will have to prove this equality, if it is to be applied anywhere.

2. The triple strictly continued cut leads to a situation in which we have generated three plane cuts through a body which are pairwise perpendicular to each other. A long tradition in speech and thought has made this figure familiar, disregarding, as it were, the body which carries it inside as cut surfaces, namely as "imagined" planes, which carry a spatial coordinate system (normally called "three-dimensional"). (It is not necessary to repeat the arguments here that a reasonable usage of coordinates is first made possible by such a "bearer," that is, by a determinate form of three pairwise orthogonal planes; in contrast, to mere counting or calculation. Likewise, the arguments that in practice, e.g., for the specification of spatial coordinates for bodies in physics or engineering, such a coordinate-bearer cannot be generated by measurement alone, need not be repeated. Only after form determinations have been sufficiently worked out, so that even the "straightness" of the measuring rod and the directions in which it must be laid or used have been established in reality, say as the edge of the laboratory table, can the localization of bodies through measurement and, with that, the application of analytical geometry occur.)

As already dealt with in Part I, the specification of three coordinate-bearing forms has often played an important role in the discussion of three-dimensionality. But the result of the historical investigation was that it was either *merely asserted* that there can be *only three* planes or three straight lines (or straight rods, too) perpendicular to each other, without being able to provide a successful proof for this. Or one simply took this *assumption* for granted, as in the assertion that *three* coordinates *suffice*, or that *exactly three* are necessary, to specify the position of a body in space. The suggested definitions for spatial terms given here in Part II (based on technical processes for the production of spatial forms by cuts through real bodies), obviously has led us to the foundation of spatial determinations which is actually used in physics and engineering. This might be enough of a reference to application for the moment, in order to now be able to clarify the role of the triad in reference to the technically and terminologically mastered form of the body assembled from eight equal right vertexes. The number three occurs in the following

> *Basic theorem of dimension theory:*
> Every simple body can be divided by strictly continued halving cuts exactly three times.

This basic theorem, which, thus, precedes a definition of the term "dimension," breaks down logically into two sub-assertions:
1. There are *at least three* strictly continued halving cuts of a body.
2. There is *not a fourth* strictly continued halving cut of a body.

This proposition shall now be proven. (The reader who is not interested in this proof can skip to page 166 below.) Comments, supplementary remarks and discussions of possible objections are strewn throughout this proof. These parts, which do not directly belong to the proof, will be set apart by square brackets so that the chain of proof remains clearly visible and can be read coherently.

The first sub-assertion is hardly a problem. The definitions of the plane surface, the right edge and right corner are already proof enough that a body *can* be divided three times in a plane cut in such a way that only equal parts result. As already mentioned above, it must first be shown in fact that only equal parts are created by each of the three cuts, thus, that the two plane surfaces, the four right edges and the eight right corners are all equal. We will deal with the proof of these three statements in a cut of its own (Chapter Nine, p. 183), because it plays an important epistemological role beyond the proof of the basic theorem, and will be decisive for the explanation of the intuitive certainty of the layman regarding spatial or dimension-theoretical states of affairs. We want to emphasize that we are merely anticipating these proofs here.

[When the principle of methodical order is not accepted, or more generally, when one sees neither the difference between action and talk about action, nor the character of corporeal properties which are artificially brought about as the result of action, the possibility of three strictly continued halving cuts could perhaps be questioned. The question might be how one knows that there are right corners, or that one can cut a body up into eight right corners? What form of experience or natural law assures us of this?

Questions of this kind are only possible, however, when one overlooks that our knowledge of space, whether everyday, mathematical or physical, is always in reference to the artificial production of spatial forms upon bodies. We would have no reliable language about spatial states of affairs if we could not count upon the basic ability of humans to move and work bodies. The mooring of the definitions which are given above to the action of cutting, and the reference which is gained in this way to familiar everyday practice, bring the doubter in the precarious situation of having to doubt his own basic abilities and to doubt that people can reach semantic agreement about such basic abilities and their results. It is neither a matter of physical theory nor scientific knowledge of the world, but is rather a bit of practical life which children of school age learn. Whoever had not successfully gone through this learning process, and would not, indeed, assume the understandability of the words learned in this process, could not even begin to formulate questions, or to express a doubt, about how or why or whether there are plane surfaces, right edges and corners in the sense of the definitions above. So we can assume that the "existence" of these spatial forms is sufficiently insured by the operational definition of these terms. The "possibility" of actions is not shown linguistically, but by carrying out actions.

It is not surprising that the first sub-assertion of the basic theorem can be proven to be true, essentially by pointing to definitions which have already been furnished. This is, indeed, also the assumption - which nobody seriously doubts - that space has "at least" three dimensions. We know of numerous examples for this triad, in science, as in everyday life, proceeding from the practice of measuring bodies like thread or wire according to length, bodies like a sheet of paper according to surface, and bodies like metal bars according to volume. The geometrization of the world in which we live, through the production of instruments, houses and entire cities has allowed us to take the distinction between length, breadth and height for granted. If, however, one wishes to give *reasons*, beyond this simple fact, as to why at least three specifications are necessary for the unequivocal localization of a body in space, then a sophisticated form-theoretical lemma must be brought in, according to which the eight right corners are equal to each other. (This is true in spite of the fact that oblique coordinate systems can be used in physics and geometry.) The extent to which the basic theorem can be derived solely from a form-theoretical access to spatial statements will be discussed in the next chapter, p. 202.]

The second, more difficult, sub-assertion of the basic theorem still remains to be proven. This is the one treated by many authors as allegedly proven or self-evident, or in some other way true and valid, without their having delivered any proof of it. So how can one demonstrate that it is impossible for a body which has been halved three times in the strictly continued manner, to be halved a fourth time in this way?

It must be remembered that halving through a restriction to the oriented fit of the originating parts can only be achieved through *plane* cuts. With this in mind, the question just posed can be reformulated to read: can a plane cut be made through a body which has been thrice divided by a strictly continued halving cut in such a way that it halves the body a fourth time, i.e., so that only equal parts are created?

The proof of the proposition that this is impossible will take the following course: The equality of all eight right corners will be accepted as a *premise*, the justification of which will, itself, later be proven. It therefore suffices to consider a single corner or right vertex concerning the question about how a plane cut can pass through a right vertex. A demonstrably *complete listing of all cut possibilities* (a total of ten), then exhausts the number of ways for the continuation of these plane cuts through the remaining seven vertexes. This will show that in *all possible* cases, *at least one* of the vertexes is *not met* by the plane cut. Thus, some of the eight vertexes, which are equal in form to each other, are divided, but at least one is not divided, which results in the *inequality* of the resulting parts. To put it more briefly and intuitively: there is no possible plane cut through eight right vertexes which would result in sixteen parts equal in form. Let us now proceed to the details of the proof.

As was already done in the proof of the first sub-assertion of the basic theorem, we must once again repeat that the eight corners (or vertexes) of the thrice (strictly continued) halved body are equal. But if every vertex is replaceable with any other one, then the same cut must be possible through every corner, or vertex. Therefore, it suffices to consider *one* corner, or vertex, and to use it to make a list of "all possible" plane cuts.

Of course, only those properties of the right corner, or right vertex, which have been definitionally established may play a role here. In other words, only those spatial forms which are present in the same way for every corner come into question: the right vertex is a part of a simple body. It bears three plane cuts, which intersect in three right edges and form a right corner. According to the definition of the right corner, or right vertex, this list of spatial characteristics is *complete*. For precisely this reason, the completeness of a list of *all possible plane cuts* can be shown with one right vertex.

Among the relevant properties of the right corner by which the number of possibilities of combining plane cuts is drastically limited, is the equality of the three plane bounding surfaces and the three right edges. That is, if one takes a right vertex from a thrice (strictly continued) halved body, one can replace it in three different ways.

This is the direct consequence of the equality of all right edges, which itself likewise requires proof as a premise, and will be discussed later (again, in the next chapter, p. 192). As long as consideration is taken of coherencies such that in reference to *one* particular edge, the two surfaces which form this edge are indistinguishable from the third, then what is true for one edge, is true for every edge. And the analogy is also true for surfaces.

In order to make the list of possible cuts and the completeness of the list clearer, a particular group of cuts will be discussed first, in regard to their ability to accomplish a fourth halving cut. One can call this first group "non-actual," or "virtual" cuts of the right corner. We are discussing the possibilities of plane cuts of right vertexes with regard to the whole, triply halved body. *This* is also divided by such cuts, which only "touch" the corner, or vertex, under consideration. There are three possibilities of this type:

1. A touching cut of this kind can simply be the repetition of a cut which has already been made, i.e., can coincide with a surface which has already been produced. It is "non-actual," in reference to the vertex in question, because it does not sever any part of this vertex; beyond this, it accomplishes a halving without a doubt, since after such a cut, all parts of the original body are equal; but it is *not a fourth* halving cut, since it only leads to eight parts, and thus does not really raise the number of parts which were achieved after the third halving cut. It cannot, therefore, deliver a counter-example for the basic theorem.

2. Another "virtual" cut is one which coincides with an edge but no surface, and which does not divide the vertex in question. Such a cut will certainly meet and divide other parts of the body, thus it is already impossible that this cut be a halving cut. From originally equal parts, namely, the eight right vertexes of the body, it divides some and does not divide others, so that after this cut all the parts can no longer be equal.

3. If the right vertex in consideration is only met in the corner point itself, without the cut coinciding with a surface or an edge, then the same argument as in case 2 is true here. This also cannot be considered a halving cut.

With that, it has been shown that the non-actual cuts, which one can regard as a borderline case of the following cuts to be discussed, can*not* achieve a fourth strictly continued halving cut.

A deficit of the previous terminology became apparent in the argument just presented. The word "corner" is applied for the *form* of a part-body which came into being through a triple cutting. On the other hand, it is needed (as in case 3) for the "corner point," which is understandable in ordinary language, but which has not yet been terminologically provided for. Since, in the following, we will distinguish between a plane cut which goes through the "corner," or "vertex," at any place, and one which exactly meets the "corner point," we now want to raise the word "corner point" likewise to the status of a term. The connection of the word "point" with "corner" shall be sufficiently protected from the misunderstanding that this already has to do with some term of geometry, or with an expression of an ideal or idealized mathematical language.

We want to proceed analogously with the edge, for which a further term, "edge line," will be established. As will be proven in the following, it is not necessary to explicitly define the terms "corner point" and "edge line" according to the rules of the art of definition of mathematics and logic. But on the other hand, they are not superfluous when it comes to expressing, briefly but clearly, that a plane cut goes through a right vertex at exactly the corner point, or exactly through the edge line. This is a more special form of cut than a plane cut which goes through right corners or edges at any place. It is, however, helpful to hold on to the following terminological connections.

Simple bodies are cut up into part-bodies which are called "wedges" or "vertexes," by strictly continued plane cuts. The form of a wedge is called "edge," and the form of a vertex is called "corner." Especially distinguished forms are the right edge or the right corner. The form of the edge is formed by two (in this case, plane) surfaces; the form of the corner is formed by three or more. The edge line belongs to both cut surfaces of the edge; the corner point belongs to all of the cut surfaces of the corner. For this reason, one could say that the edge line bounds both of its cut surfaces; the corner point likewise bounds all edge lines which come together in it.

After the three non-actual plane cuts of a right corner (i.e., the plane cuts which only coincide with the corner point, an edge line or a surface) have been excluded, it becomes quite easy to afford an overview of all combinationally possible cuts (referring to the corner point, or the number of the edge lines or surfaces which are met). As an illustration, a special case is taken, whose specialty however only serves a clear spatial imagery, but not the line of proof. One thinks of a sphere, which is cut up into eight equal sectors by three plane cuts, pairwise perpendicular to each other, which meet in the mid-point. One now considers such a sector, and allows the plane cuts to wander outward, so to speak, proceeding radially from the corner point.

The most general case of a real cut, i.e., one which splits the sector into two parts, meets all three surfaces and all three edges. The cut through all three surfaces now moves outward in such a way that it meets, in sequence, three, two, one or no edges. These are a total of four cases. (Cf. Fig. 9).

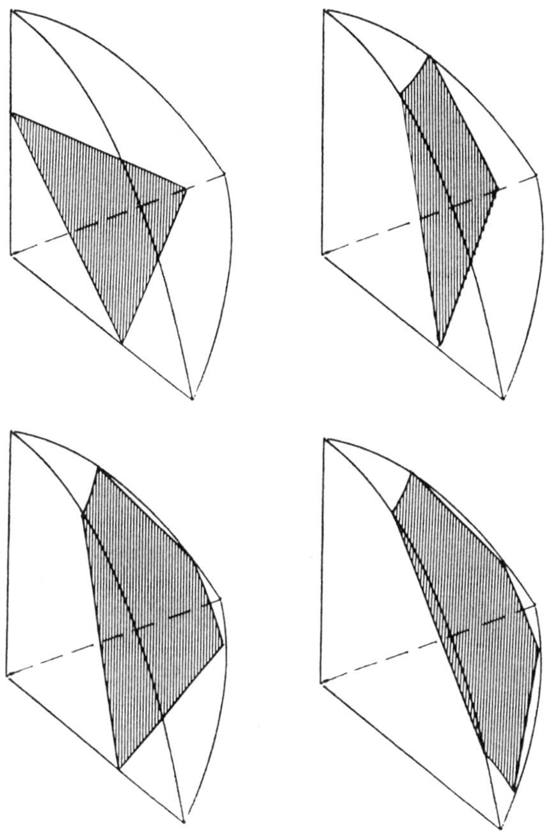

Figure 9. Plane cuts through three plane surfaces of a right corner, and through 3, 2, 1 or 0 edges.

The next cuts to be considered are those which only divide two of the three surfaces. Obviously, of the four combinational possibilities of meeting three, two, one or no edges, the first two do not come into question, since three (different, of course) edges are, according to definition, formed by at least three (different) surfaces. This is also true for two different edges; three different surfaces are also necessary for their formation. But since, according to the condition, only two different surfaces are to be divided, only cuts that meet one edge (namely, the one they share) or none at all, come into question. The cut through two surfaces of the sector provides a total of two possibilities. (Cf. Fig. 10.)

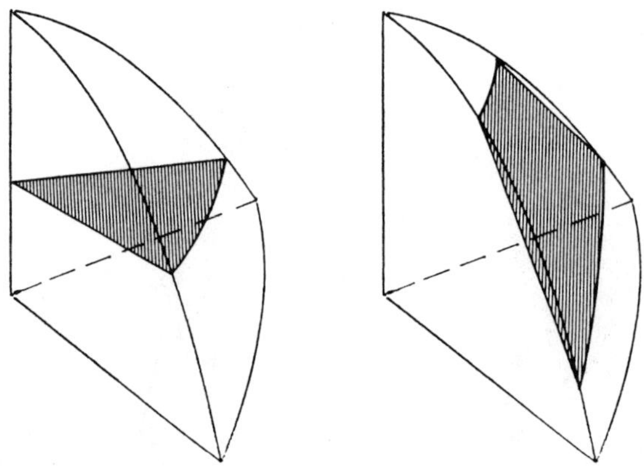

Figure 10 Plane cuts through two plane surfaces of a right corner, and 1 or 0 edges.

Because of the definitional connection between edge and surface, there is no possibility of a cut that meets an edge in the next case of a plane cut, which only divides a single plane surface. Only the case of the cut that does not meet any edge remains as a possibility here. (Cf. Fig. 11.)

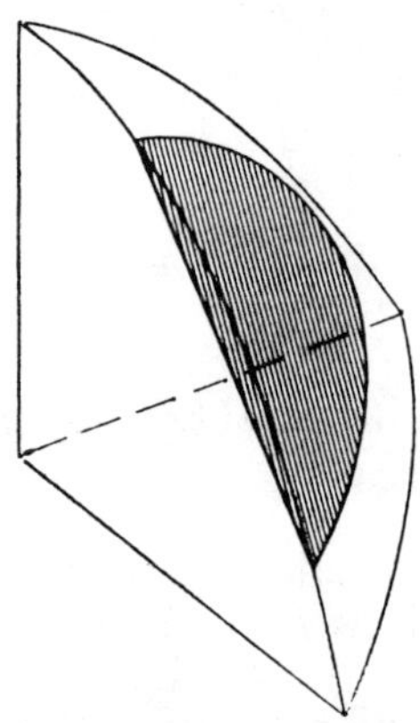

Figure 11 Plane cut through one plane surface of a right corner.

Finally, in regard to the number of plane surfaces of a sector which are divided by a plane cut, the case can arise in which none of the plane surfaces are met. Of course, then no edge will be met, so this again adds a further, singular case to the series which has already been listed. (Cf. Fig. 12).

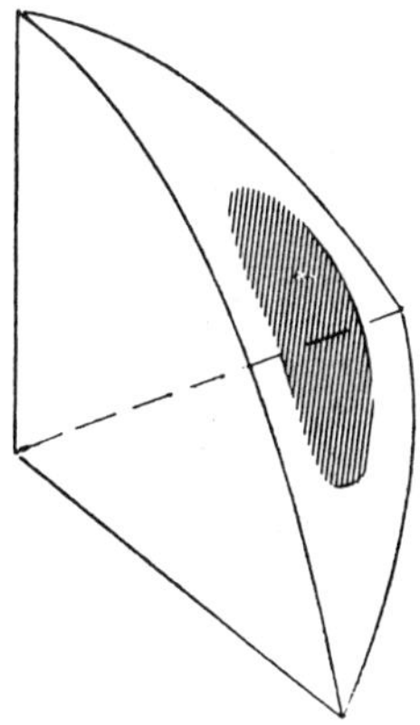

Figure 12 *Plane cut through a right corner without meeting a plane surface.*

With that, we have gained eight possible cuts through the sector up to now, which cover all combinational possibilities of three surfaces and three edges in consideration of the definitional connection between surface and edge - with an important restriction: the corner point itself has not yet been taken into account in the list. Obviously, plane cuts through the corner point are a special case, which differs from the previous cut possibilities, since per definition, a plane cut meets all the edge lines which come together in the corner point, but does not divide these. Hence, even a plane cut through the corner point cannot really divide any of the three edge lines, i.e., cannot cut through them in such a way that two parts of the edge line would result.

Every layman will be able to immediately discover that only two sorts of (real, i.e., cutting the right vertex into two parts) plane cuts through the corner point are possible. The one sort of cut will coincide with an edge line, and will actually divide the opposite surface somewhere. (Cf. Fig. 13). The other sort of cut will not completely coincide with any edge line and leads to two vertexes, one of which is formed by three surfaces, and the other by four. (Cf. Fig. 14). However, since a proof is supposed to be presented here, one must ask why these two cases represent a complete list of all possibilities.

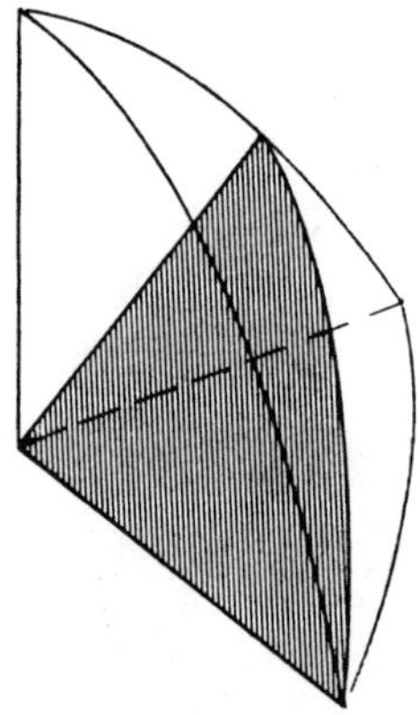

Figure 13 *Plane cut through one edge and the corner point.*

It follows from the definition of the right corner that each pair of the three edges has a plane surface in common. Therefore, no plane cut which coincides completely with two edge lines comes into consideration; since these would be non-actual cuts, and it has already been shown that they cannot accomplish a fourth strictly continued halving cut. It follows *a fortiori* from this that no plane cut can coincide completely with all three edge lines. Hence, of the four combinational possibilities (the plane cut coinciding with three, two, one or not any edge lines), only the last two cases remain.

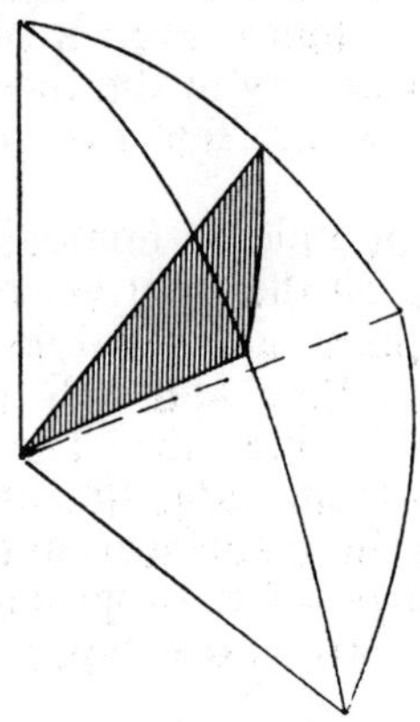

Figure 14 *Plane cut through the corner point alone.*

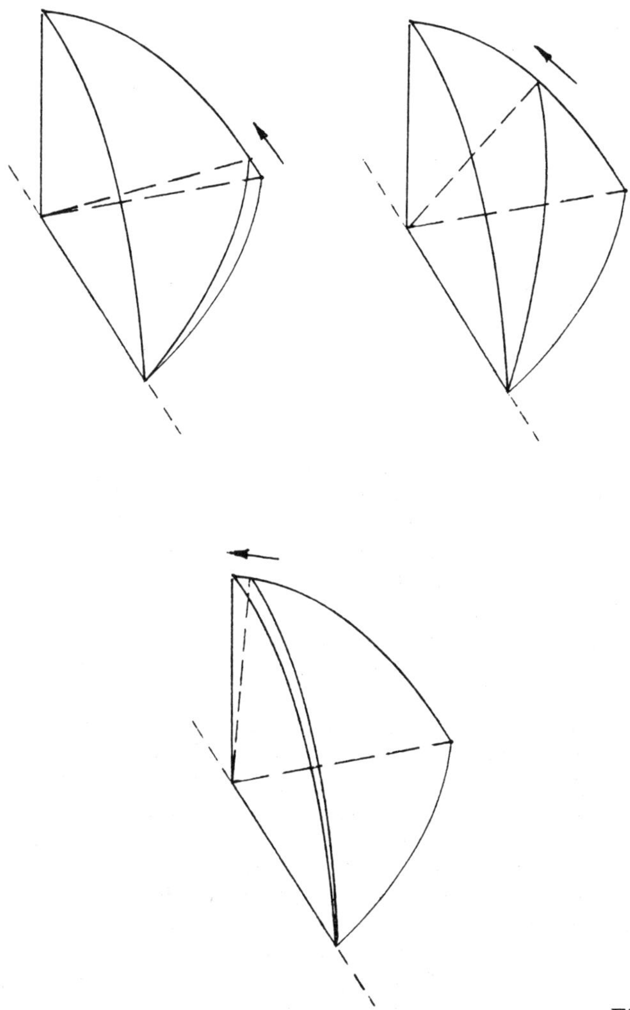

Figure 15

Obviously these arguments differ from those of the previous eight cut possibilities, which did not cut through the corner point. There the surfaces were considered, here it is the edge lines. Therefore, a further consideration is presented here to show that, with these two cases, *all possibilities* of cutting the three surfaces of the right corner, with the additional condition that the plane cut goes *through the corner point*, are now, in fact, accounted for. The set of all cuts which

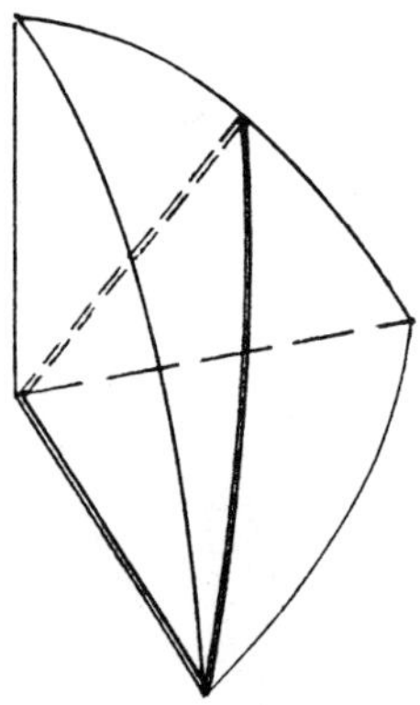

Figure 16

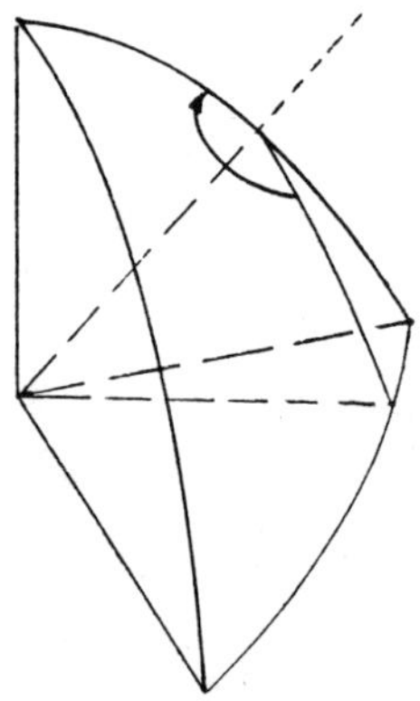

Figure 17

coincide completely with an edge line E (Cf. Fig. 15), and which really divide the vertex, lies between the two virtual cut surfaces S_1 and S_2 which form this edge line. The cutting plane is, so to speak, rotated from one bounding surface to the other around their common edge line. In this way all possible straight lines on the opposite surface S_3 are generated by the corner point. In other words, there is not a plane cut through the corner point which really divides the surface S_3 - which lies opposite the initial edge line E - which has not already been included in the list.

Per definition, every one of these real cuts divides the right vertex into two parts. For any given one of these divisions, the newly generated pair of edge lines *E'*, *E"* (Cf. Fig. 16), can now be divided, again (for the right corner which is assembled by fitting both parts together), by a plane cut in such a way that this completely coincides with the pair of edge lines *E'*, *E*." Intuitively, the cutting plane rotates 180 degrees around the chosen rotation axis. (Cf. Fig. 17). Since all of these cuts go through the corner point, all possible cuts through the two other opposite surfaces S_1 and S_2 are accounted for. So, regarding the three surfaces S_1, S_2 and S_3 there is not a possible cut through the corner point that is not covered by both cases - cuts that completely coincide with an edge line, and cuts that do not completely coincide with an edge line.

There are, thus, a total of *ten real possibilities* of dividing a right corner with a plane cut. These ten types of cuts are once again presented in the following figures. The numeration of the types corresponds to the list given here: the first eight types are cuts that do not go through the corner point, the last two cuts do. The first four types cut through all three surfaces and, in so doing, also cut three, two, one or no edges. Types five and six cut two surfaces, and one or no edges. Type seven cuts a surface and no edges, and, finally, type eight cuts neither a surface, nor an edge. The last two types go through the corner point, and with type nine, cut three surfaces (by coinciding with an edge line), with type ten, only two. (Cf. Fig. 18).

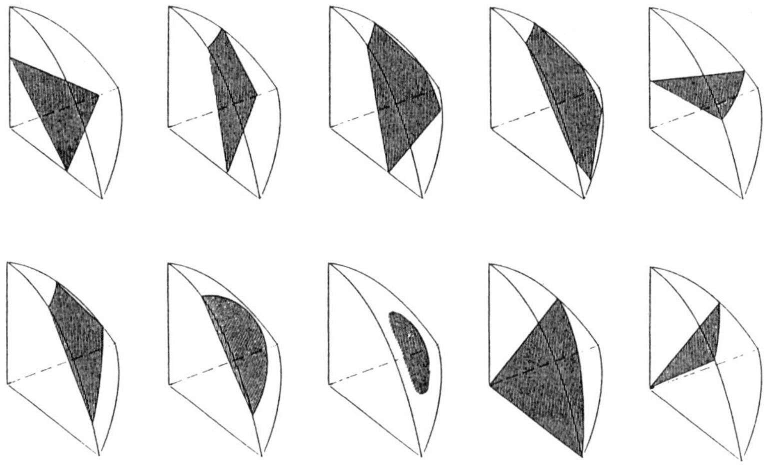

Figure 18. The ten possible plane cuts through a right corner.

In these figures, a sphere sector is considered for purposes of illustration, just as in the text. This restriction is, however, not essential. A sufficient condition for the logical completeness of the list of ten cut types, is that the original body is simple in the sense of the definition given above (Cf. p. 144).

Now that a complete overview is available of the possibilities which exist for a plane cut through a right corner, it remains to be shown for these ten types of cuts that none of them lead to a fourth strictly continued halving cut. In other words, we have to discuss how each of these cuts through a corner continue through the other seven corners of the original body. In order to accomplish this in a stringently logical manner, a few new words must be agreed upon for better clarity:

Vertexes (corners) which *only* touch at the corner points are called "*opposite*"; vertexes (corners) which *only* touch at the edge line are called "*adjacent*"; vertexes (corners) which touch at the one surface are called "*neighboring.*"

In order to enable a clear identification of the eight different vertexes, their surfaces and edge lines, we will introduce designations: the three halving cuts through a body generate three edge lines which are pairwise perpendicular to each other - these will be numbered proceeding from their common cut point.

Edge segments on different corners which lie in a straight continuation to each other, hence, are formed by the same pair of cuts, receive the same index, but are distinguished from each other by plus or minus signs. In accordance with the customary representation using Cartesian coordinates, we will speak of positive and negative axes, and will use these as a reference for a listing of the eight corners or vertexes, as is given in Figure 19:

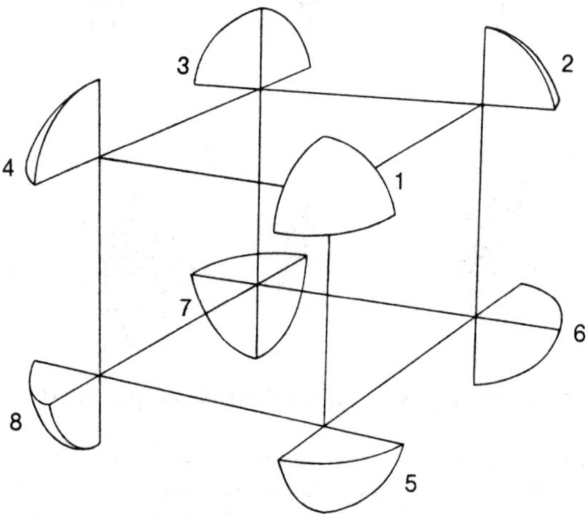

Figure 19

Accordingly, the corners (vertexes) are formed by the respective half-axes:

Corner (vertex):	Half-axes:
1	1, 2, 3
2	1, 3, -2
3	1, -2, -3
4	1, -3, 2
5	-1, 2, 3
6	-1, 3, -2
7	-1, -2, -3
8	-1, -3, 2

For a clearer graphic representation, in Figure 20 we have dispensed with marking the edges which result from halving cuts *on the outer surface* of the original body. Instead, only the three axis pairs which represent the result of the triple strictly continued halving cut have been specified. With these the ten possible cuts through the Corner 1 will now be marked according to the following convention:

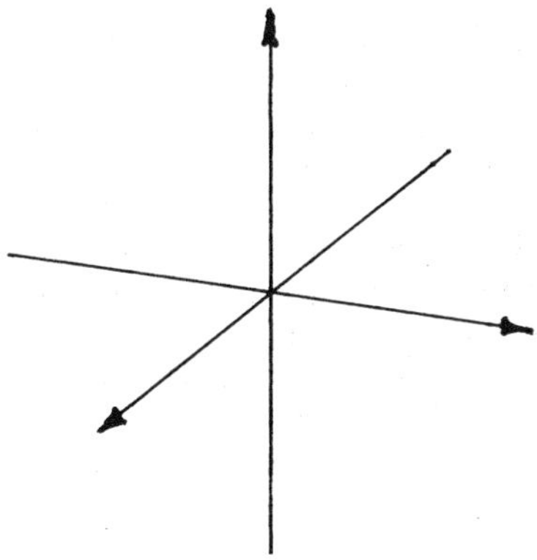

Figure 20

A newly formed edge (as a cut through a surface of Corner 1) will be represented by a straight dashed line. Arrows on the end of these dashed lines mark the exit of the dashed line from the outer surface of the original body, assembled from the eight right vertexes. A small circular spot will be drawn in wherever a cut meets an edge of the right vertex. The cut corners and, respectively, the corners that are not cut, stand in a row beside the single figures, ordered according to the criteria of neighboring (letter *n*), adjacent (letter *a*) and opposite (letter *o*). The symbol *x* means "cut," and the symbol *x'* means "not cut."

Cut Type 1:

	x	x'
n	2, 4, 5	
a	3, 6, 8	
o		7

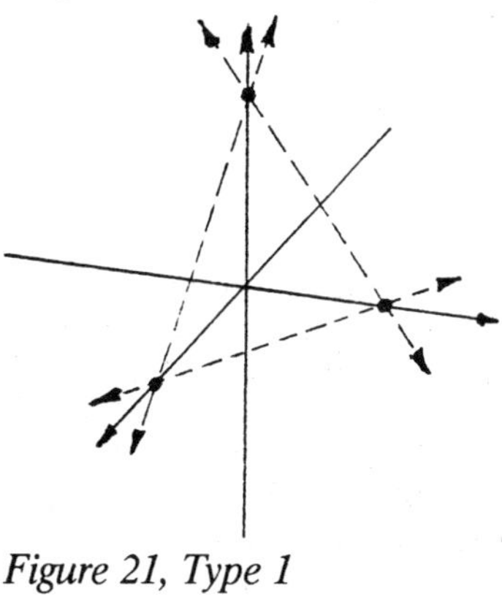

Figure 21, Type 1

Cut Type 2:

	x	x'
n	2, 4, 5	
a	6, 8	3
o		7

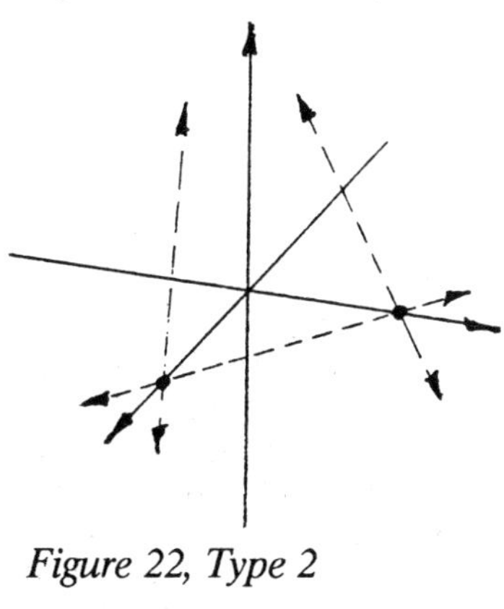

Figure 22, Type 2

For the cut of Type 2, however, the relation of the cut in reference to the other seven corners is not clear. A cut of Type 2 only demands, per definition, that three surfaces and two edges are cut on Corner 1. Regarding the inclination of the cutting plane in reference to the uncut third edge, it is only established that this may not be altered in such a way that this third edge is indeed met. All other inclinations, such as proceeding from a fixed cut line through both cut edges 2 and 3, are compatible with the definition of cut Type 2. With that, such cuts come into consideration which meet the "extension" of the uncut edge on Corners 5 and 7. This leads to a second variation for the cut of Type 2:

Cut Type 2, 2nd Variation:

	x	x'
n	2, 4, 5	
a	6, 8	3
o	7	

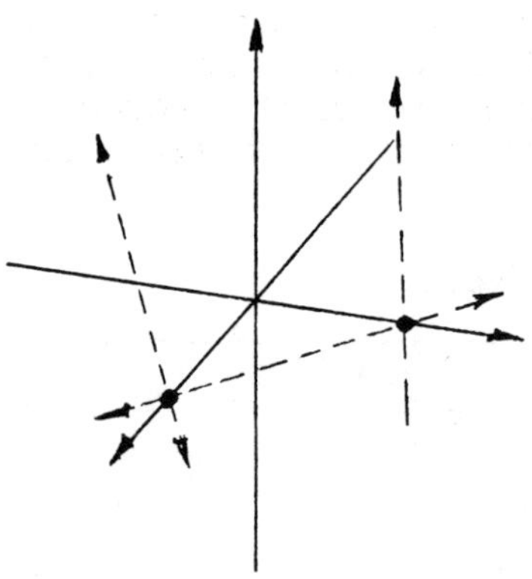

Figure 23, Type 2

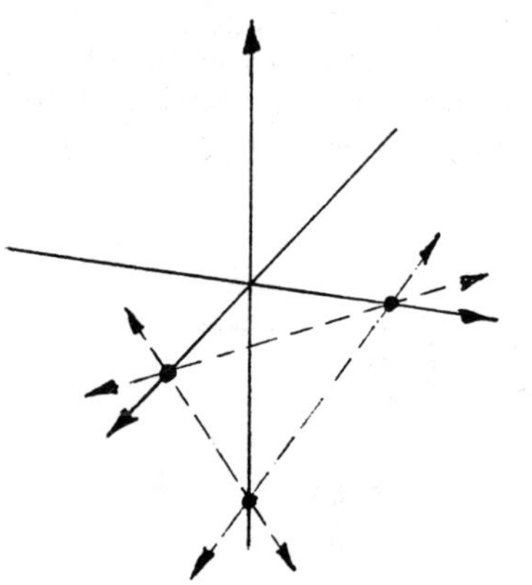

Type 2

A glance at the last figure, however, immediately shows that this variation of the cut of Type 2 for Corner 1 is undistinguishable from the cut of Type 1, to the extent that one refers this now to Corner 7. In other words, the second variation of the cut of Type 2 through Corner 1 is undistinguishable from a Type 1 cut through Corner 7. But we already know that a cut of this type leaves at least one of the other corners, namely exactly the opposite, uncut. Hence, we will refrain from discussing this second variation of the Type 2 cut - this case is already covered by the Type 1 cut.

Cut Type 3:

	x	x'
n	2, 4, 5	
a	8	3, 6
o		7

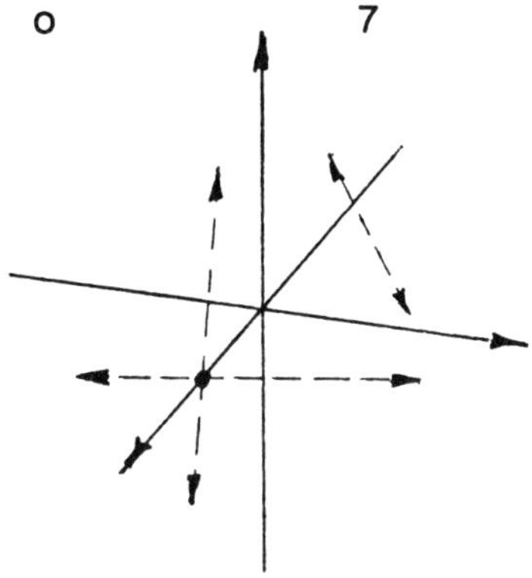

Figure 24, Type 3

Cut Type 4:

	x	x'
n	2, 4, 5	
a		3, 6, 8
o		7

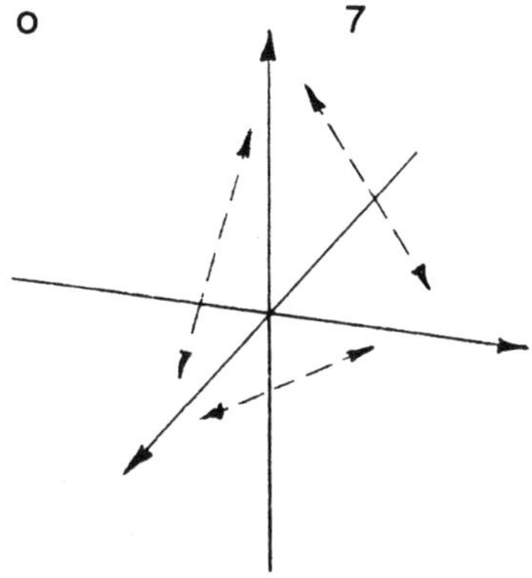

Figure 25, Type 4

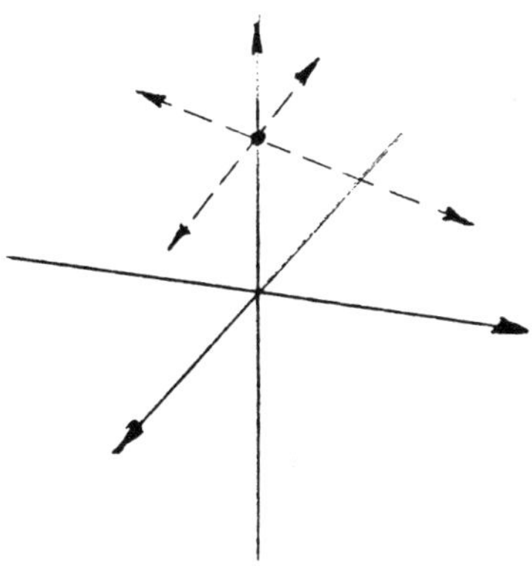

Figure 26, Type 5

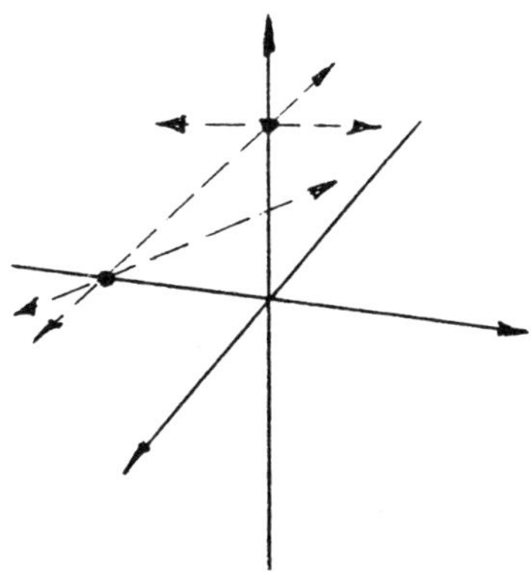

Type 5

This yields an ambiguity which is similar to that of the Type 2 cut: the condition for the Type 5 cut, which demands that two surfaces and one edge of Corner 1 are to be cut, is compatible with various inclinations of the (new) cutting plane in relation to the uncut surface (2, 3), as for a fixed point of intersection on edge 1. In this case, either none, one or both edges - 2 and 3 - will be met on Corner 3.

However, just as above, these cases shall not be listed and treated as proper variations, since, in this manner, cuts of Types 1 or 2 are obviously obtained on Corner 3. These have already been treated, and it has especially been shown that they always leave at least one of the other corners uncut. We will therefore now proceed to Type 6 cuts:

Cut Type 5:

	x	x'
n	2, 4	5
a	3	6, 8
o		7

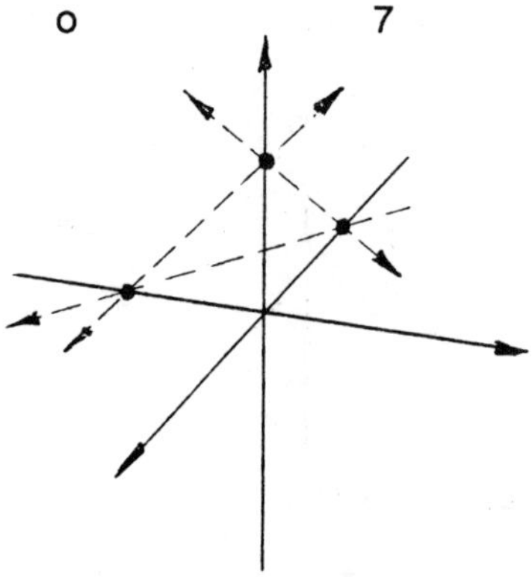

Figure 27, Type 5

Cut Type 6:

	x	x'
n	2, 4	5
a		3, 6, 8
o		7

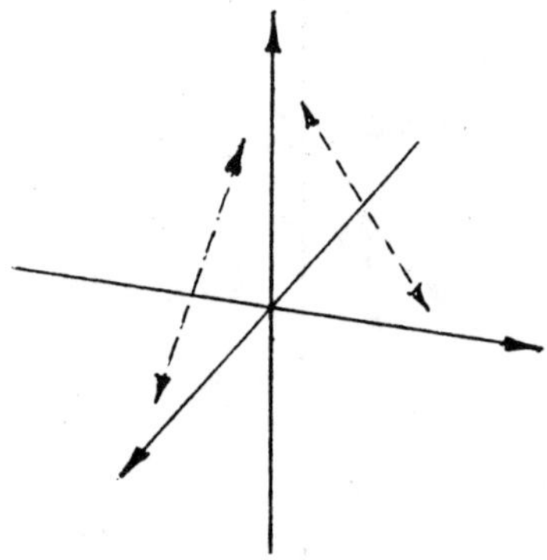

Figure 28, Type 6

Cut Type 7:

	x	x'
n	4	2, 5
a		3, 6, 8
o		7

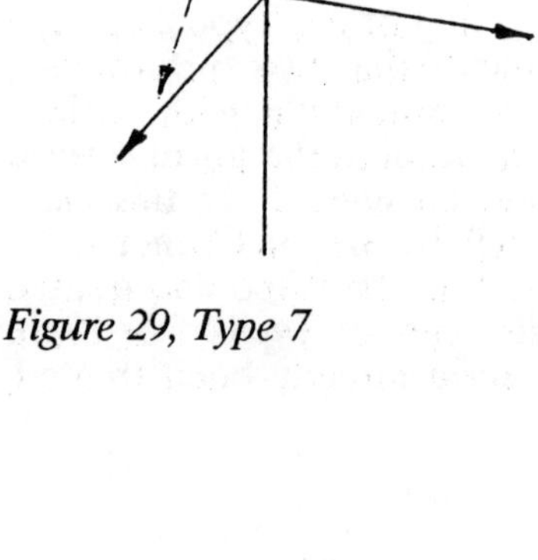

Figure 29, Type 7

Cut Type 8:

	x	x'
n		2, 4, 5
a		3, 6, 8
o		7

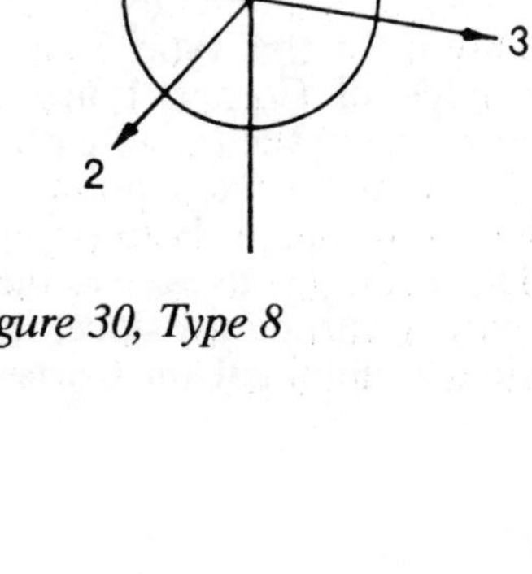

Figure 30, Type 8

Since the new cutting planes go through the corner point in cut Types 9 and 10, the position of the newly created edges cannot be represented clearly in the same way as the first eight cut types were represented. Additional helping lines or surfaces are therefore drawn in. In the case of Type 9, these are perpendiculars from any given point on the newly created edge to the original plane cuts. These perpendiculars should show how the position of the new cut edge is to be understood.

Cut Type 9:

	x	x'
n	2	4, 5
a	8	3, 6
o	7	

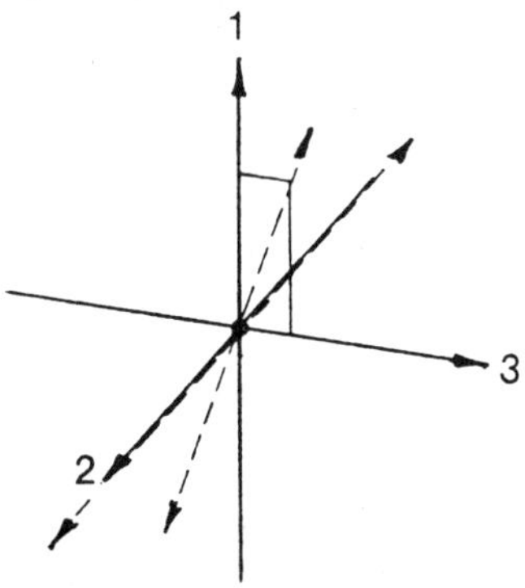

Figure 31, Type 9

Cut Type 10:

	x	x'
n	2, 4	5
a	6, 8	3
o	7	

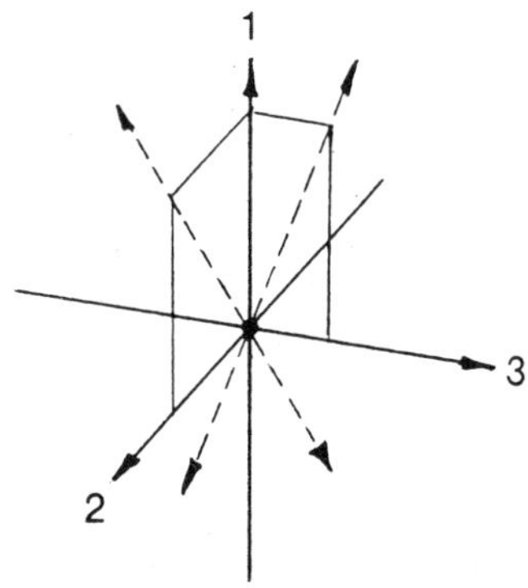

Figure 32, Type 10

In the figure for the Type 10 cut, two pairs of these perpendiculars are necessary; every newly created edge goes through the corner point, and their position to the original three cutting planes is determined in this way.

As a result, the continuation of the ten possible plane cuts through a corner reveals that at least one of the other corners remains uncut. Since it was demanded of a strictly continued halving cut that - in the literal sense of the word "halving" - *all* the created parts are equal, and since all eight parts were equal before the fourth cut, it follows that no plane cut through a thrice (strictly continued) halved body can be a fourth strictly continued halving cut, *which was to be demonstrated*.

The same state of affairs can also be intuitively grasped as follows: when a simple body is divided by three plane cuts which are pairwise perpendicular to each other, there is not a plane cut which makes sixteen equal parts from the eight right corners of this body. With that, the basic theorem of the dimension theory has been proven. The form and content of this proof require some explanation and a discussion of possible objections:

The basic theorem of the dimension theory must not be understood as meaning, in letter or spirit, that no *division* of a body (and derived from this, no division "of space") into more than eight equal parts exists. One can already read in Euclid the proof that there are regular icosahedra (20-sided bodies). If one connects the mid-point of the icosahedron to each one of the corner points with a straight line segment, one obtains twenty equally formed pyramids with triangular bases. Equality of form is to be understood, in this context, in the same sense as replaceability for plane surfaces, right edges and right corners. But *this cutting up* of a body is obviously *not primarily accessible in method*, but is first achievable through the concepts of plane, straight and orthogonal, i.e., through the processes of production for plane surfaces, right edges and right corners. In other words, there cannot be a cutting process which cuts up a natural body which is not defined in its form into twenty pyramids of equal form. Only with the help of a geometry which is already successful - to which the concept of congruence must then be added, and consequently also those of the circle or the sphere - can one produce equal three-sided pyramids of an appropriate edge ratio such that twenty of them can be assembled into a whole icosahedron.

Of course, it still has not been explicitly shown why the triple strictly continued halving cut of bodies should have any meaning at all for the everyday or scientific understanding of concept of dimension. After all, the term "dimension" has not yet been defined. Before this next step is made, a few considerations will be brought forward concerning the sense in which the basic theorem can be considered *proven* by the ten cut possibilities listed above and their continuation through the entire body.

In the strict sense of the word, usual in mathematics, a *proof* is the logical deduction of an assertion from given premises. To this end, it has proven itself useful in logic and mathematics to introduce a formal language of their own, which not only allows for abbreviation and a clearer presentation than a formulation in plain language; the (claimed) explicit definition of the applied symbols also assures that no silent premises are applied in the steps of the proof. Regarding the form of the proof, it could be objected that this is not a proof in a strict, logical sense because it does not offer a formalized logical deduction in application to a theory which has already been expressed in a symbolic language. There are three replies to this objection:

1. The logical stringency of a proof does not stand or fall based on a presentation in symbolic form. It is a mistaken opinion - which has

even become an irritating fashion in some branches of science - that a matter is more exact, the more it is presented in a symbolic language of formulas. Even with logically correct deductions in a formula presentation, a lot of room for interpretation can remain regarding the results, when the leap to formal language, from the realm of objects which are being spoken of, is not sufficiently bridged by definitions. David Hilbert makes it understandable, in view of the most current history of mathematics and the turning of mathematics to a formalistic self-conception, that a great definitional deficit must be accepted in the linking of the realm of objects with formal language, but it is nevertheless not systematically justified.

This state of affairs also exists in regard to the dimension problem, for which mathematics has developed a subdiscipline of its own, namely, topology. The self-conception of the mathematicians in reference to the content and conduct of their own discipline has, indeed, led to brilliant mathematical considerations, as well as definitions for the term "dimension." However, at the same time, this has blocked off any access to the question concerning the sense in which the empirical space of everyday life and engineering is three-dimensional.

Finally, it seems to have seeped clear into popular opinion that the precision, hence the reliability, of a theory or line of argument depends on the degree of its translation into a formal language. The extensive scientific debate about the meaning of physical technical terms, for instance, shows that when it comes to applying the mathematical theories to real things and events in the laboratory, there is a lassitude in the highly developed mathematical apparatus of theoretical physics that a layman would never imagine. The concentration on the mathematical theories of physics which is found in the theory of science has, for a long time, ignored its historical side and even today does not do justice to its technical/practical foundation. The fact that the mutual definitional dependence of individual expressions within a theory might be far less significant for the validity of this theory than the constitution of its basic concepts in technical laboratory practice, is to this day almost inaccessible for most theoreticians in physics and the theory of science.

2. The proof given above of the basic theorem of the dimension theory could certainly be formally presented, too. If one wished to resort to something familiar here, then a methodical access to analytical geometry would first have to be opened. Then the example of a sphere which is split up by three pairwise orthogonal planes could be used to arithmetically treat the cut possibilities of a fourth plane through the sphere sectors, case by case. Such a proof, which still works with elementary mathematical means (of approximately the high school level), must justify in the first step why the sphere, or

in a more general case, an asymmetrical ellipsoid, is chosen as the representative of a simple body. To this end, a ("topological") equivalence of the sphere, or the ellipsoid, to all simple, i.e., convex bodies must be proven, which would then certainly require a mathematics which is no longer elementary, but is rather demanding, indeed, problematic, regarding its own premises, especially the concept of dimension. What is missing is precisely the link between the (topological) theory and speech about real bodies, which is embedded in practical interaction.

Another possibility for a formal proof would be possible if speech concerning plane cuts and their results were formalized. One would then have to formulate and prove lemmas like one which states that a plane cut which really divides an edge or a right corner cannot cut the straight continuation of this edge on the neighboring corner. With the help of such lemmas, one could sort out from all the combinational possibilities, according to which the eight right corners can be cut in ten different ways, the set of possibilities which are compatible with the conditions of the relative position of these corners and the planeness of the cut.

This formally intricate possibility for a proof, which is only accessible to a reader familiar with formal languages, has been dispensed with in order to demonstrate with an example the possibility of a proof which is non-formal, clearly presented, intuitively understandable, and methodically admissible. This has the epistemological advantage of starting with the pre- and extra-scientific practice of dealing with bodies, in this way remaining free from the conceptually stronger requirements of the individual sciences, in particular mathematics and physics. This independence shall also serve the claim in the next cut of making the intuitive certainty of the layman regarding the three-dimensionality of space understandable.

The objection could be made against the proof given above, that in the discussion of the ten cuts of Corner 1 in the other seven corners of the body, it is merely *asserted* that some corners are cut and others are not cut. At first glance, this seems to be either a mere exhibition, utilizing a functioning concept of space on the part of the author and reader, or a report of actually performed cutting experiments. In other words, one could have the impression that it is either a matter of empirical sentences or of sentences requiring proof. Hence, the mere exhibition could come too short, in the same sense as the exhibition of the arithmetical sentence 2 x 3 = 6 by counting a set of apples by twice taking three apples out of a bushel. Does this proof not make use of sub-assertions which, for their part, would also require proof?

To this point, let us consider again two essential steps of the proof: the first step consists of a listing of the possibilities of a plane cut through a corner, with the claim that this list is conclusively complete. The second step consists of showing that none of these possible ten possible cuts can truly divide all the other seven right corners of the body, so that every possible plane cut leads to divided as well as undivided - hence, unequal - parts of the original body.

What could still be objected to in the first step? Could it still contain unclarified premises? The proof brought forward here depends upon the completeness of the description of a right corner through the determinants of three plane cuts and three right edges, as well as the stipulation that the original body be simple. It is then a trivial fact that all the combinational possibilities (cut through the corner point vs. cut not through the corner point; cut through three, two, one or no surfaces, each with a cut through three, two, one or no edges) were considered, and reproducible beyond a doubt. The only premise that requires proof here is the indistinguishability of the three edges and the three surfaces. As already stated, this point will be addressed later.

It could be asked, however, in regard to the geometrical axiom which states that exactly one straight line is determined by two different points, how one knows that a plane cut cannot meet a straight edge without completely coinciding with it. The principle of methodical order, however, prohibits such a question as inadmissible. The straight edge (and, with it, the "straight line") was, in fact, defined by the plane cut. So this is not a case of two undefined basic concepts like "plane" and "straight" standing next to each other, but is rather the context of the production of spatial forms on bodies which provides a definitional context that excludes the described possibility of a double cut of a straight edge by a plane (without the straight line and the plane coinciding).

But how does it stand with the continuation of the ten cuts in the second step of the proof? It is a matter of fact that the following knowledge was applied without any debate: if a cut through Vertex 1 really divides one of its surfaces, then the fitting surface of the neighboring vertex, which lies against it, will also be divided. This state of affairs could certainly be made into a provable sentence through a further elaboration of the definitions of cutting, plane, the reassembly of the body from its parts (created by cuts), etc. But no additional clarity or reliability would be gained by this. Possibilities for formal differentiations should not be taken as goals in themselves, or as a type of occupational therapy. Whoever would doubt that a real division of a surface results in the real division of the fitting surface of a neighboring vertex which lies against it, no longer partakes in the linguistic convention which is familiar to everyone, and which is fixed in the practice of cutting up bodies.

It was likewise not discussed in detail here, how the continuation of the real division of edges which lie next to each other results from the continuation of the real division of surfaces which neighbor each other. Finally, for the sake of completeness, it must be considered how it can be stepwise proven that in, say, the case of a Type 1 cut (not through the corner point; three surfaces and three edges are cut) the corner opposite the cut corner is not met.

It is possible to perform all of these steps in the proof. They can be carried out by any layman as a type of busy-work in logic at a very elementary level. However it would not be epistemologically informative. Especially important - and decisive for the thesis of this book - is that *a silent assumption of three-dimensionality* has not been smuggled into this proof *anywhere*. Therefore, we will dispense with giving reasons for individual, uncut corners, beyond those already given in the tables on pages 162 to 166. The proof of the basic theorem of the dimension theory will also stand up to logical examination in its present form.

Instead, we now turn to the important remaining tasks of this book: a definition for the term "dimension" is still to come, and the reliability of the proposal given here for the layman as well as for the technician and physicist must be shown, and the epistemological status of the approach presented here must be clarified.

CHAPTER NINE

The Spatial Concept of Dimension and Its Universality

THE DEFINITION OF "DIMENSION" AND THE THREE-DIMENSIONALITY OF THE BODY

The word "dimension" is one of those scientific technical expressions which has found a wide and diverse use in ordinary language. Does a language usage, scientific or extra-scientific, exist which would suggest a *methodically correct definition*?

When, in ordinary language, it is confirmed that an object has a "new dimension" or "another dimension" as opposed to some other thing, this means, in the most general sense, a conspicuous difference, but mostly it refers to a more or less exactly nameable difference in magnitude. Whether this concerns the costs or measurements of a construction project, the time expenditure or the work force involved, technical risks or political obstacles, "dimension" occurs in connection with an estimation of magnitude for almost any chosen criterion. The original Latin meaning of the word in its loosest form can be found here: the prefix "*di*" or "*dis*" means "separate" or "different," and the second part of the word means measure or measurement. Thus, the literal translation of "*dimensio*" means *spatial measurement in different (or all possible) directions*.

The very vague usage in ordinary language compares with usages which are likewise very different in the sciences; physics and mathematics, for instance, apply the term "dimension" very differently. In analytical geometry, the dimension of a space is specified by the number of coordinates which are necessary for an unequivocal determination of a point. On a line, this is one number; on a surface, two; in ordinary space, three; and generally, in n-dimensional space, n coordinates are needed. Illustrations of these definitions might make use of spatial forms, like the lines in a drawing plane on a body, but the mathematical definitions make no use of these, but rather aim at precisely the independence from such illustrations.

The physicists speak of dimension in the sense of a type or a designation of a physical magnitude. A minimal condition for the truth of any physical equation which compares, for instance, magnitudes before and after an event (such as the kinetic energy of two billiard balls before and after impact), is that both sides of the equation bear the same designations. In other words, a physical equation always formulates the *equality of magnitude of homogeneous quantities*. A language usage is obviously already at work here which is derived from the geometrical-spatial usage, according to which one must think of a mechanical equation with the magnitudes path, time and mass, for example, as being represented by various coordinate axes.

This metaphorical way of speaking in physics can then be used for any "measurable" quality, independent of whether or not it has all the logical properties of a scale, such as can be defined for the spatial extension. Thus, whether magnitude ratios or only magnitude differentiations can be compared, or whether in the end only a arrangement of discrete elements is defined, whenever a realm of objects is considered which can at least be consecutively numbered, "dimension" can be spoken of.

The Latin word "*dimensio*" does not correspond to any literal translation in Greek. In Greek, there is on one hand the word "*diametros*" (διαμετρος) for the diameter of the circle or sphere, and on the other hand, the word "*diastasis*" (διαστασις), which means something like "distance," "standing apart" or "division." Hence, the difference in meaning between the words *dimensio* and *diastasis* is accounted for, on one hand, by the fact that the prefix of the Latin word means the variety or complete listing of the directions, while the prefix of the Greek word is more in the sense of "through"; besides which, the Greek word is more general than the Latin word to the extent that *measurement* is not yet mentioned. More elementary than measurement is *dividing*, which represents a fundamental concept in the ancient Greek view. That which must be *traversed* once, twice or thrice according to the *Latin* sense of the word is, in the *Greek* view, singly, doubly or triply *divisible*. This difference between the basic operations, as it were, with whose help the difference between line, surface and body is determined, can offer an alternative as to where a systematic modern definition must start. Should one begin with *measurement*, as suggested by modern quantitative natural science and its mathematical means, or is there a more elementary, more exact candidate, which *precedes* measurement *in method*, for a definition of the concept of dimension?

The discussion of the historical approach in Part I has shown that all "higher," i.e., not purely spatial, attempts at providing reasons for three-dimensionality, already make use of spatial concepts. It is to be shown that such methodical dependencies affect the choice of the starting point for the definition of "dimension."

The "analytical" arguments, beginning with laws of force and ascending to the logical consequences of entire physical theories, are already disqualified as definitional beginnings. In these, geometry is applied right from the beginning, through a highly developed art of measurement, and use is made of the spatial differences of dimensions. The mathematical treatment of the dimension problem in topology would suggest defining a dimension as a continuum, and seeing the realization of the continuum in motion, for example - as is attempted in the approach of Clifford. Does it therefore recommend itself to bind the definition of the concept of dimension to the motion possibilities of a body in space?

In theoretical physics textbooks, where the means of description for the motions of bodies are brought together, point-shaped bodies are considered first. These can only perform translation motions. More complicated bodies or figures (formed of two or more points) can, additionally, exhibit rotations. Of course, here the talk of the translation or rotation of a body or figure is always a relative description within coordinate systems. In the *technical realization*, this means that relative motions are considered between two *bodies* on which certain point-shaped places are marked. However, one must obviously already know that the point-shaped marks on the bodies, which represent the coordinate system, must not all lie in one line or even in one plane, but that *at least four different points* must generate at least three different spatial directions. In other words, any talk about the relative motion of two bodies to each other already requires form-theoretical concepts like "plane" or "straight" for the identification of translations and rotations. These considerations are not only true for motions, but for any relative position as well. Thus, motion and position do not come into question as primary methodical accesses to a definition of "dimension."

The fact that such a starting point is, finally, not to be found in the magnitude of a body has likewise already been discussed in the historical context. So linking the definition of dimension to the meaning of the Latin *"dimensio"* does not recommend itself, because for every (length) measurement, a spatial direction, and a material bearer for it, must already be form-theoretically distinguished. With that, only the *form of bodies* remains as a *methodical start for the definition of dimension*.

What is meant in form theory, when it is stated that a body is three-dimensional? Of course, a language usage is intended here which is determined for any given body (even natural, unworked bodies of any given form). We therefore propose a definition which is closely connected with the literal sense of the Greek word *diastasis*:

A right corner is removed from a body *B*, which has been halved (strictly continued) three times, so that we obtain a "hollow space," such as is formed by any normal corner in a room. Any place on the outer surface of any given body B' can be brought into contact with one of the three surfaces of such a hollow corner. One now shifts the body *B'*, while maintaining its contact place on the first surface, in such a way that a contact place is achieved on the second surface and, finally, also on the third. *B'* then lies in the corner in such a way that it has at least one point of contact with each of the three, pairwise perpendicular, surfaces. One now lays three right vertexes, with one edge each fitting one of the three respective hollow right edges, and then slides each of these up to body *B'* in such a way that all three vertexes have at least one point of contact with *B'*. In this way, *B'* is *enclosed*, as it were, *in a container made up of six plane surfaces, such that it has at least one point of contact with each of these six walls*. (Cf. Fig. 33).

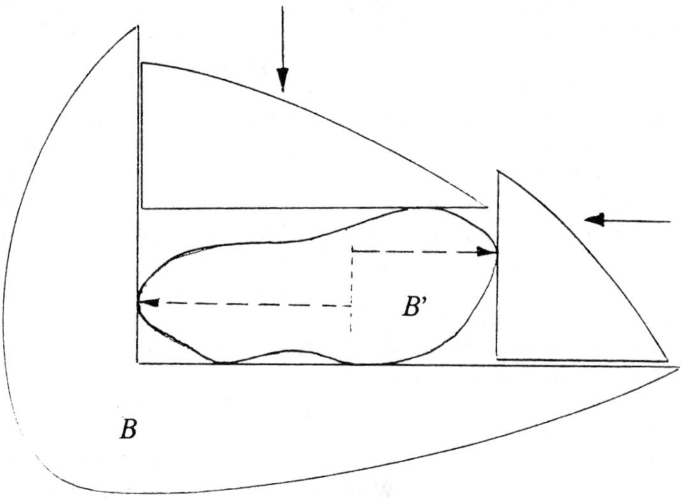

Figure 33

If we name opposite surfaces parallel - without any geometrical reservations at first (they will be discussed later) - then a *procedure* has been specified in which, proceeding from a triply strictly-continued halved body *B*, a body *B'*, in any given form, can be enclosed in a container of three pairs of parallel surfaces. In the sense of a preliminary definition of the concept, we call each *distance between these enclosing parallel surfaces "dimension."* The procedure presented here insures that any given body has exactly three dimensions, since every body can - at least in principle - be enclosed in a container by this procedure.

The (still preliminary) form-theoretical definition of "dimension" raises a host of questions, which must now be clarified in preparation for the final definition.

If the triad of dimensions of bodies is traced back to the triad of parallel surface pairs with which these bodies can be completely enclosed, the objection could be raised that further enclosing pairs of parallel surfaces are, indeed, additionally possible. For example, pairs of parallel surfaces could again take the place of the spatially opposite pairs of edges of the original, cuboid enclosing container (the edges would be "broken"), and six further pairs of enclosing parallel surfaces would be created in this way. Hence, the body *B'* in question would have ten dimensions in the sense of the definition. Complementary to this, it could also be asked why *at least* three pairs of enclosing parallel surfaces are necessary.

These questions or objections challenge whether the definition of dimension using the enclosure of bodies within pairs of parallel surfaces truly shows that *minimally and maximally three* dimensions are to be assumed for bodies.

Until now, only the translation of the Greek word diastasis has been given as a motive for the proposal to understand the dimension of a body as the distance between two parallel surfaces which touch the body. This is certainly more a historical reason than a convincing argument. Argumentative reasons first arise in the further context of spatial statements about bodies, and this is not restricted to a definition of the word "dimension" as an end in itself, as it were. Picking up on the pre-understandings which were already explicitly mentioned in the first chapter, we can name the description of form, magnitude, position and possibly the motion paths of bodies, as the goal of spatial statements about bodies. (Of course, the choice of means for this "description" depends, again, on the practical context in which the description is needed.)

Our common pre-understanding suffices fully here, according to which we make a sufficiently good specification for the position and magnitude of a body, in many contexts, when we say that it lies in a (cuboid) box, and therefore has a smaller volume. The procedure specified above, of enclosing the body in question in a minimal container (minimal relative to the arbitrarily chosen first point of contact) by laying it in a hollow right corner, and then sliding three further right corners along the edges of the hollow corner, fulfills precisely the purpose of giving quite a good specification about the position of the body, and a rough but reliable first estimation of the volume. It becomes obvious from this that it is the complete enclosure that matters here. If only two pairs of parallel enclosing surfaces were chosen, such that a "channel" with a rectangular cross-section would result, then the body would be shiftable within it, and no reliable specification about position or magnitude could be gained. With that, the minimum number of three enclosing pairs which are, according to the requirements, pairwise perpendicular, has been proven.

But what is the case with the maximum number? It would obviously improve the estimation of the volume if ever more pairs of parallel surfaces were shoved up against the body until contact points were achieved. In this way, more and more of the hollow spaces between the body and container would be cut off from the original cuboid enclosing container. Thus, there must be another reason for seeing three as the maximum number.

The answer, of course, is that three surface pairs *suffice to form a complete enclosure* of the body. Of course, one can also skeptically inquire how this is known. An objection of this type, which mathematical theories really must take seriously, could be stated as follows: The generation procedure for a minimal container for a body, which consists of sliding three vertexes against it, assumes, for reasons as yet unexplained, that new edges arise. Of interest here are the three edges which are formed pairwise by the three vertexes, and are, thus, not generated by the original hollow corner. How does one know that a cut is at all attained? Could it not be, instead, that any two or three new contact planes approach each other asymptotically as it were, without forming a cut edge? (Cf. Fig. 34.)

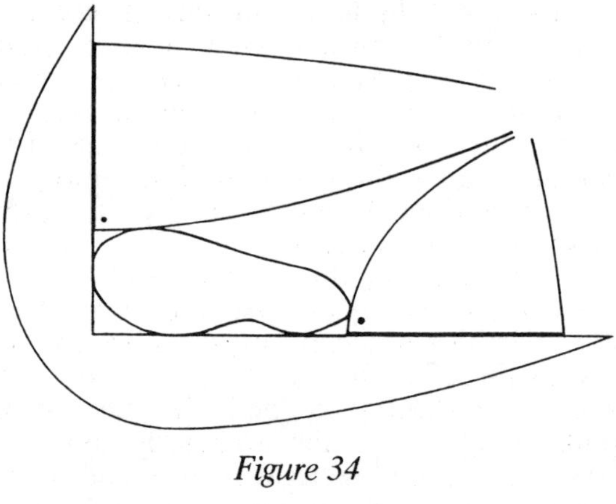

Figure 34

In such a case, a complete enclosure would not be achieved.

But in connection with our production procedure this cannot be an objection. To shove three right vertexes against body *B*' only serves to substantiate the assertion that for every contact point on *B*', a second, "opposite" one can be achieved on a contact plane parallel to the first. The procedure was not, however, intended to prove the thesis that humans can actually succeed in producing *closed containers*, in particular, cuboid ones. In other words, "closed" is not defined by the procedure described in the definition of "dimension."

Instead, directions can be given, with the available vocabulary, as to how any simple body can be divided by six plane cuts in such a way that the first two cuts form a right edge, and every further cut forms a right corner. In this way a part-body (from the inside of the body) is created, which is only bound by plane cut surfaces. If one takes this part out from the inside, and then lays the other parts back together again, one receives, by definition, a *closed container*, such as is required for the considerations presented here. With that it has been shown that three pairs of parallel cut surfaces suffice to generate a closed container. Questions or objections regarding whether it is actually shown by the dimension definition using the complete enclosure of bodies, that *bodies have minimally and maximally three dimensions*, have, therefore, now been sufficiently answered.

A question of an entirely different kind concerns the *universality* of the proposed *definition*: is it true for "all" bodies? One motive for this question could be that we cannot, of course, submit bodies of any given magnitude to an enclosure procedure. In such cases, it is traditionally favored to say that the enclosure possibility of bodies in cuboid containers is "principally" valid. But which principles come into question here? What kind of knowledge is it which is exhibited in the assertion that something can be done which nobody can do (e.g., to pack the moon in a carton)? Is it empirical knowledge, principle empirical knowledge, definitional knowledge or knowledge of an entirely different kind, which allows one to assert such principle possibilities?

The answer is very simple, even if it is not allowed for as a possible answer in many traditional philosophies. That which is "principle" about the enclosure procedure, lies in *knowing the procedure*. We would not, in an everyday context, deny that someone knows the procedure for swimming or for playing the flute, simply because no water or flute is available at the time. That is, we would not accept the lack of suitable circumstances for realizing a procedure as a reason for denying that someone knows the procedure. Indeed, in our everyday understanding, we go even further: even if the circumstances are suitable, and the action scheme is not actualized, we do not see this as a reason to dispute knowledge of the scheme. It is, for example, highly improbable that the reader of these lines has ever counted so far that the number sequence 987654321, 987654322 occurred. Nevertheless, one would not see this as an occasion to assert that the reader does not know how to get to the second number from the first.

So if the principle possibility of completely enclosing any given body using the described procedure is claimed, then this is meant in the same sense. The knowledge of action schemes does not lose its many definitional consequences or its primary importance for method when we, as a rule, dispense with actually carrying out the enclosure of bodies, or when we know that we lack the technical means for especially big bodies and always will. The universality of the definition of dimension as regards the entire world of bodies is thus assured by knowing one of the appropriate procedures.

A further group of questions is more concerned with the geometrical background of the given definition. The mathematician and the physicist know that coordinate systems for the localization of bodies are not dependent on the *orthogonality of the coordinate axes* (and the layman can easily visualize this). Of course, every point of the drawing surface can be reached with a parallelogram construction when both coordinate axes include an oblique angle. What role does the orthogonality of the edges and corners play for the definition of dimension?

The answer does not concern a mathematical, but rather an epistemological, aspect of this approach. Plane cuts through simple bodies were defined in such a way that a set of directions can be stipulated in primitive teaching and learning situations through a simple recipe. The right edges and the right corners as forms which are achieved through strictly continued division are distinguished to the extent that they are *reproducible free of prototypes*. One needs no particular prototype for a distinguished oblique angle. The instruction to produce equal parts with cuts suffices. (We will not go into the fact that plane cuts through right wedges which halve angles likewise lead to equal parts, and that every angle is ultimately reproducible free of prototypes through the continued angle halvings and the assembly of suitable parts. In the end, this procedure only makes use of the fact that the first application produces a right edge. Oblique coordinate systems are produced in the same way as orthogonal ones - only with more effort.)

Enclosing containers in which the edges are not rectangular lead to a certain ambiguity. The corners of an oblique parallelepiped (i.e., a container whose six surfaces are formed by parallelograms) are no longer equivalent in reference to the choice of the original hollow corner. One need only consider the two "maximal" possibilities: a corner is either formed from three acute, or three obtuse, angles. It is self-evident that the resulting contact points on body *B*' will be different, depending upon which possibility is chosen. These inequivalencies are even more grave when the edges that come together in the original corner do not include the same angle. Then the points of contact on body *B*' will even depend on which of the three boundary surfaces first touches *B*', and which touches it next.

Oblique containers from three pairs of parallel boundary surfaces also completely enclose the body *B*', but they are more complicated and "come later" in reference to methodically ordered reproducibility and, regarding an unequivocal determination of contact points, they achieve less. Therefore, a cuboid container is to be preferred.

The question concerning the orthogonality of the edges or corners of the enclosing container must not be confused with the question about whether the given definition of "dimension" is true *only for Euclidean geometry*, or if it is also true for non-Euclidean geometries. When opposite boundary surfaces were called "parallel" in an earlier passage of the text, this was "without any geometrical reservations at first." But such reservations are now appropriate, and must be discussed. To this end, one considers the original hollow corner and a right vertex which is shoved along it on one of its edges. In this way, a pair of parallel surfaces is generated. "Parallelism" is defined, in this, by the condition of having a common perpendicular - namely, the edge line, along which the right vertex is shoved. This definition itself, however, does not assure "Euclidicity," i.e., the truth of the parallel axiom or a proposition equivalent to it.

To clarify the question of whether the definition of dimension is Euclidean or not, let us consider the proposition about the orthogonality of the fourth corner of the rectangle, as a proposition which is equivalent to the parallel axiom. If a rectangle is constructed in a plane, one erects two perpendiculars P_1 and P_2 on a given straight line L and, following this, one erects on P_1 a perpendicular P_3 which intersects P_2. The sentence that P_2 and P_3 intersect each other is equivalent to the parallel axiom. This "plane case" has its counterpart in the generation of the enclosing container for a three-dimensional body:

The original hollow corner has three edges. Each of the three right vertexes which are shoved along these edges generates two more edges a piece, so that, by definitional stipulation, there are a total of nine edge lines, and nine angles which are included by them. Only four corner points have been determined until now (namely, that of the hollow corner, as well as those of the three vertexes in the hollow edges). What results from this are the cuts of the three moved vertexes among themselves, which lead to three new edges and four new corner points. As one can easily imagine, nothing is determined for the angles which result from this. In other words, each of the six boundary surfaces of the enclosing container originates from the construction of three right angles, which are thus stipulated, and contains a fourth angle which "results from the construction." Whether this fourth angle is a right angle in each case determines whether or not the construction on a whole is Euclidean - and therefore, whether the suggested definition of dimension is bound to the Euclidicity of geometry.

We will consciously leave this question open for the time being, with a reference to the later section, "Concerning the Certainty of the Physicist." The procedure for producing an enclosing container is not affected by this. Right edges and right corners are defined, the production procedure for enclosing containers has been described, and the *definition of dimension is, therefore, applicable to Euclidean as well as non-Euclidean cases*. (Of course, this can only mean that bodies are, by definition, three-dimensional, regardless of whether one talks about their other properties with the means of Euclidean geometry or of a non-Euclidean geometry, and regardless of which reasons determine the choice of a geometry. To express the matter rather loosely in ordinary language: even if non-Euclidean, bodies are still three-dimensional.)

In conclusion, let us also consider the question of whether the concept of dimension which is determined through the enclosure of bodies in a minimal cuboid is at least compatible, if not fundamental, for other concepts of dimension, in which, for example, the *structure of continuity* or the *measurability of magnitudes* play special roles. The "distance" between two parallel enclosing surfaces is, of course, formed by the edge line of the minimal cuboid, proceeding from the original hollow corner and its corner point as a common point of reference. The (topological) structure of these edge lines has not been discussed until now. But the enclosure procedure, in which a shifting of bodies relative to each other occurs, makes precisely these edge lines appear to be continua. When edges and corners are seen as realizations of geometrical lines and corners, then it makes sense to say, for every relative position of one of the moved vertexes to the hollow corner, that the corner of the vertex divides the edge line. And according to the argumentation which originated from Aristotle, between any two different points of division there are always further such points, i.e., the edge lines are always divisible into, once again, divisibles, i.e., they are continuous. The mathematical definition of dimension in topology as continuous sets is therefore thoroughly compatible with the proposal presented here. Of course, this continuum definition is not strictly accomplished here, but then this is also not necessary in order to prove the three-dimensionality of bodies.

In accordance with common language usage, the measurability of "distance" is associated with length. This understanding likewise does justice to the definition of dimension given here, as long as a procedure is only *additionally* proposed as to how one can compare the edge lengths of the minimal cuboid and arrive at a ratio for their relative lengths. The procedure, which was already described by Euclid, of alternating subtraction (ανταναιρεσις) achieves such a definition of the magnitude ratio. In other words, a supplementation of the definition presented here, which corresponds to the Greek *diastasis* (by definition, measurement as the determination of length ratios), now also leads to an adequate interpretation of the Latin *dimensio*. (In this, the "*dimensio*" is the measure of the edge length of the enclosing container). These need not, of course, correspond to the lengths of the straight lines between the points of contact which originate on opposite enclosing surfaces. Because this connecting line must, or course stand perpendicular to the enclosing surfaces. (Cf. Fig. 33 on p. 175).

To bring the obvious questions and objections concerning the presented definition of dimension to an end, let us take a look back at the beginning of the chapter. In a little linguistic criticism, talk of "space" was terminologically neutralized as talk of spatial distinctions. It now becomes apparent that stripping this traditional philosophical problem of some metaphysical obscurity was not particularly costly. Whether space is understood as an ordering scheme for things, or as empty space in the sense of a position or motion possibility of bodies, the "three-dimensionality of space" has become understandable. As long as those space-filling bodies are thought of whose "space-filling" consists of the fact that two bodies cannot be in the same place at the same time, the three-dimensionality of space has been explicitly defined. All of the relative positions of the bodies to each other are also perfectly mastered in a technically reproducible and controllable manner. If, in contrast, one thinks of "empty space," then the three-dimensionality of an artificially produced hollow space or container is also explicitly defined.

Thus, the methodical beginning of spatial speech with artificially produced forms obviously accomplishes that which is to be expected of a critical linguistic revision of traditional space philosophies. The approach does justice to every aspect of traditional spatial concepts in a terminologically clarified manner: empty space and filled space, the qualitative and the quantified, continuous and metrical space. Finally, even talk about relative spaces is accessible (through specifications of relative positions for hollow spaces); only "absolute space" is excluded here. But since the conception of absolute space is tied to a cinematic or a dynamic description - motion or acceleration relative to absolute space should be able to be sensibly described or recognizable - a definition of absolute space does also not belong to a "purely spatial" theory of space.

If one asks about which "properties of the outer world" are claimed for this definition of dimension and the thesis of the three-dimensionality of space, or about what else the definition and thesis could perhaps be based on, it is the elementary, mundane practice of working bodies. This can be arranged normatively in such a way that certain forms (followed by their equal forms) are technically reproducible. Thus the basis for the definition and thesis about three-dimensionality is finally a *capacity for action*, which, in fact, provides for an active interaction with bodies as they simply are "by nature." But this capacity for action is not, itself, dependent on any (empirical, for instance) theory about the natural properties of bodies. It is much rather the case that all that we know about the workability of bodies, we know from the activity itself. If we did not know it in this way, we would not know it at all - not even physics could make this knowledge available. The fact that the truth of the thesis of three-dimensionality is, therefore, based upon the historically proven worth of a particular practice of craftsmanship and not upon some purely verbal definitional trick, will be more exactly proven in the last three sections of this chapter.

Now here is the *final definition* according to these clarifications:
"Dimension is the distance between two opposite plane (body outer) surfaces, which stand perpendicular to a third plane surface."

("Dimension" is defined for hollow spaces and for filled spaces (body), and this is for spaces of any given form, as well as for rectangular spaces.)

Dimensions stand perpendicular to each other.

According to the basic theorem of the dimension theory, every hollow and filled space has exactly three dimensions.

UNEQUIVOCALITY AND PROTOTYPE-FREE REPRODUCIBILITY

In the proof of the basic theorem of the dimension theory, use was made of the proposition that all right corners, all right edges, indeed, even all plane surfaces, may be treated equally. The discussion of the cut possibilities of a single right corner as part of a simple body was also supposed to provide information about the cut possibilities of the other seven corners of a triply (strictly continued) halved body. The proof of this assumption shall now be dealt with, but not only in order to close up a gap in a logical argument which is still open. Rather, the problems which are raised here are of great epistemological significance, since they will finally provide the reason why the layman, as well as the scientific specialist, is convinced, with great intuitive certainty, of the three-dimensionality of empirical space; why we, indeed, hold three-dimensionality to be "necessary" in a sense which has yet to be clarified.

As has been shown in Part I, there have already been many different approaches to explaining three-dimensionality. This book is not intended to merely add a further theory to these, but to go beyond all the older approaches with the claim of making the special "necessity" or "reliability" of the assumption of the three-dimensionality of space. This clarification must span the spectrum from the most elementary everyday knowledge with little theoretical or terminological reflection, up to the most sophisticated theories and technical procedures of mathematical science.

Therefore, this section of the chapter will primarily serve to recall those elementary spatial states of affairs which are "somehow" familiar to every layman and specialist, but which have normally not been carefully discussed terminologically.

Above, we have made use of the *form equality of, respectively, all plane surfaces, right edges and right corners* among each other. This form equality is defined by replaceability within a triplet of surfaces or edges or corners - with the edges or corners an additional condition, that they be laid on a flat plane or in a right hollow corner, is required. In other words, we are dealing with a special form of universality or generality, which must now be discussed, as phenomenon and fact, before we proceed.

A plane piece of the outer surface of a body is unequivocally determined in a double sense: for the sake of distinguishing between the two we will call these unequivocalities "singular unequivocality" and "universal unequivocality." First of all, every plane outer surface is *singularly unequivocal* in the sense that (by definition) two counterparts which fit it, also fit each other. If one prepares two plaster casts, these also fit each other - perhaps with overlappings. With that an operational criterion has been gained which allows the performance of a procedure for reaching a decision in case of doubt regarding whether a particular surface is plane or not. The availability of such a procedure creates, of course, a larger independence from the competence, hence, the authority, of a mere passive observer. In other words, where such a procedure is unknown, and where two observers, looking at a piece of outer surface, are of different opinions concerning whether it is plane or not, no consensus is possible. A transition to a more intersubjective controllability of spatial statements, thus to a higher generality of truth, is first achieved with a control procedure of the type mentioned.

It is trivial, for all definitions, that the purpose of the definition lies in the application to as many cases as possible. The operational control procedure achieves an identification of all plane surface pieces. The same applies analogously for the right edge and the right corner.

The *universal unequivocality* differs from the singular unequivocality in the sense that all plane surfaces fit each other. The difference between the types of unequivocality is obvious: two surfaces S and S' are called plane when, for every fitting surface F_1 and F_2, there are respectively F_1' and F_2', so that the definition of planeness is fulfilled. The assertion that if the singular unequivocality of *S* and *S'* is given, *then S* and *S'* also fit each other, goes beyond the definition of plane and beyond singular unequivocality. It states that *all plane surfaces fit each other*, regardless of what production or control context they might be taken from.

Even the layman can easily imagine that the two unequivocalities are not already logically true based upon the definition of plane. Because the assertion of the singular unequivocality makes a leap from "some" to "all": if there are "some," for example, two, fitting pieces for a plane surface which also fit each other, it does not yet logically follow from this that all fitting pieces of a plane surface will also fit each other. But even if this singular unequivocality is accepted as certain, the universal unequivocality does not follow from it, which, again, makes a leap from "some" to "all": if it is true for all fitting pieces of "some" surfaces that these also fit each other, the formal-logical derivation cannot be gained from this , that "all" plane surfaces fit each other. (It could be that there are, so to speak, various classes of surfaces, which class-internally fulfill the definition of plane and singular unequivocality, but which, in comparison with surfaces from different classes, do not lead to a fitting.) The unequivocality can first be proven by an expansion and specification of the definition of plane in which the aspect of the fitting of two planes during a shift is taken into account.

Before we turn to this task, an epistemological digression should show which goals are followed in this. The predominance of the empirical natural sciences favors an intellectual climate in which the layman as well as the expert is inclined to approach a problem, like the three-dimensionality of space, as a search for properties of natural laws in the outer world, instead of raising this problem as a question about our cognitive conditions. Indeed, it was an organically created intellectual climate which the natural sciences inherited and which attached too little weight to reflection on the conditions of cognitivity. And yet, it is precisely this diffuse, as yet unclarified, feeling that the three-dimensionality of space is necessarily true, which can first indicate that it is precisely *not a property of the outer world*, but the *conditions of cognition* which lie in human disposition and practice which promise a solution here. And it is for this reason that this book lays so much weight - in a descriptive and analytical procedure here, to start with - on the portion available to humans, on the well-chosen ingredients of knowledge about three-dimensionality, as it were.

Of course the forms of artificially worked bodies are *not natural objects*. The technical practice of the production of plane surfaces, right lines and right corners, as well as the art of standardized speech about such forms are cultural accomplishments. It cannot, however, be the task of a philosophical reflection to merely describe these cultural accomplishments in a historical review, and to analyze them from a distance without involvement. Because even the transmission of a tradition - in this case, spatial distinctions - is, perforce, always carried out normatively. Norms (or more generally, rules) for speech and action provide the continuity which allows for the comprehensibility of speech, and the developmental capacity of theoretical, as well as practical, abilities.

Therefore, this book takes the path of carving out a normalization of speech and technical practice for the case of the three-dimensionality of space which is factually accepted, as well as well-equipped with good reasoning. First of all, it must once again be established as a simple historical fact, that the conception of the three-dimensionality of space is always bound to a *particularly distinct situational invariance*. An example should illustrate this. While we accept a wide-reaching cultural dependence in other areas of spatial distinctions, e.g., the art of measurement (the internationality of the natural sciences and of commerce has not succeeded in imposing a unified system of spatial measurement throughout the entire world), and while there are the most diverse coordinate systems, measurement procedures, spatial theories etc. for the respective contexts of application, the three-dimensionality of the body is accepted in all cultures and all languages. A situational normalization of units of measurement, for instance, does not occur. And with that, no reference to distinct standards can come into question. The plane and the right angle, even if only in the distinction between horizontal and vertical, are equally ubiquitous for all peoples. This "omnipresence" is (with reference to the word "vertical," this is in spite of the gravity of the earth) an omnipresence in human action. Plane surfaces, right edges and corners are artifacts.

It is a peculiarity of these fundamental geometrical forms, to have not experienced their circulation throughout history and the world in the form of copies, but as the dissemination of craftsmanship. The art of being able to produce the fundamental geometrical forms anew every time, in a Robinson Crusoe situation, as it were, is the actual seat of the human cultural accomplishment of geometricizing nature, of quantifying and technically mastering it. For this reason, it is there that a philosophical discussion must make its starting point, and must clarify which historically successful and reasonably grounded norms make the "necessity" that space has three dimensions comprehensible.

In reference to fundamental geometrical forms, the capability for technical action is directed at the consistent reproduction of equal spatial forms. This equality is not achieved, however, through casting, that is, through the production of copies, but through the equality of the procedure itself. (Whether the equality, indeed, the schematization of human action has it roots in early stages of technology, in the successful beginnings of agriculture, in myth or in ritual, cannot be answered here.) Our modern technological world, at any rate, would not be possible without the reproducibility of spatial forms - and this reproducibility is clearly "*prototype-free*." It does not make use of a distinct prototype, which is to be reproduced in a copy. The universality of engineering and its success is grounded rather in know-how, in engineers', technicians' and natural scientists' knowledge of procedure, which is always effective - at any time, at any place, independent of a special reference to situations or persons.

The usefulness of the prototype-free reproducibility of fundamental geometrical forms is not, of course, exhausted in the certainty of being able to repeat a procedure which was once described, in the same way that a ritual act can be repeated. The technical interest in repetition lies in the *certainty of the same results*. For the technical purpose of the spatial forming of natural bodies, it is of the greatest importance to know that the same procedures have the same results. Independent of the individual production history of artifacts, it can be expected, in light of the equality of the production procedure, that the products exhibit predictable properties in regard to each other. Bodies with plane surfaces, right edges and right corners can be stacked in a predetermined manner. Rectangular bricks baked in the brickworks, beams and boards cut in a sawmill, and rolled rails and sheet metal - building materials which were *produced separate from their direct application* - were the beginning in cultural history of a general use of this knowledge.

However, such knowledge is neither empirical knowledge in an individual psychological sense, nor in a culture-historical sense, nor in the sense of a systematic natural science. Nobody who has understood the production procedure for spatial forms in terms of its purpose, will repeat it and then supply the *empirical hypothesis* that the results would probably fit each other suitably. Instead, he would always interpret the lack of a fit as an indication of a *flawed production*. The relative properties of spatial forms to each other which are considered here, are always, and in every regard, understood as the results of actions, as action goals, in fact, which is finally due to the human practice of forming bodies. The distinction of the plane surface and right angle, as born by the theory of the three-dimensionality of space proposed here, is also based on act theory, i.e., to be understood more precisely as the logical consequence of schematic production and control acts. And with that we come back to the two unequivocalities of the fundamental forms of geometry.

The singular unequivocality (a surface is called plane when two fitting pieces fit each other) does not present a special problem as long as one considers the pragmatic context. Whoever does not wish to be restricted to merely looking at a questionable piece of surface, but also wishes to perform a reliable control procedure, will make two castings and will check their fit. (Today in industry, of course, entirely different, e.g., optical, procedures are employed; this is, however, irrelevant here since the reliability of these procedures is based on controlling the control instruments, which methodically must finally be traced back to the fitting of bodies.) In the practice, two fitting pieces normally suffice for a questionable surface. Should it be demanded for the definition of plane that all fitting pieces for the surface in question also fit each other? The practice seems to favor this. Then it is only a seeming logical paradox that the reliability of exactly one pair of fitting pieces is practically the assumption of a logical generality. If, namely, even *one* pair of control surfaces do not fit each other, one would deny the surface in question the predicate "plane," hence, one would practically draw a conclusion about all possible fitting pieces. In other words, concerning the practical implementation of controls, the definition of plane is universal in regard to the controlling fitting pieces, even if the logically correct formulation is not the "all quantor" (it obtains "for all" fitting pieces of *S* that they also fit each other), but the "singular quantor" ("there are" two fitting pieces of *S*, which also fit each other).

The universal unequivocality, however, raises greater difficulties in the sense of real burdens of proof. How can it be expected, in regard to the definition of plane, that "all given" planes, right edges and corners, correspondingly fit each other? What is arbitrary here is the individual history of production. If the same prescribed procedure is followed in *different* workshops or laboratories, can it then be known how the products of this procedure stand to each other?

Obviously this cannot be known with unlimited generality. For instance, it cannot be known beforehand, whether two plane plates which come from different workshops might not exhibit a strong chemical reaction when laid against each other and suddenly heat up. Previous knowledge can only be sought or claimed here in regard to spatial properties. Can it be known beforehand, as demanded by the universal unequivocality of the definition of plane, that all plane plates fit each other, regardless of which fitting pieces are used in the selection?

The analytical-empirical tradition in the theory of science, which began with the Vienna Circle, gives a negative answer to this question. In it, every general statement about objects in this world only has the character of an empirical hypothesis. But the reason for this lies in the fact that this analytical-empirical tradition has overlooked the active character of knowledge gain, its goal orientation and the artificiality of the examined objects of the world. Nobody would dispute that the general validity of mathematical results is derived from following rules in calculation. But what is possible when dealing with the self-generated symbols of mathematics, can also be carried over to intercourse with self-generated properties of real things. How this is to be done in the case of the universal unequivocality of the definition of plane, has already been dealt with for an operative foundation of geometry within the framework of a so-called "proto-physics," i.e., an operative foundation of the art of measurement which methodically precedes empirical physics. Only as much of this will be repeated here, as is necessary for the definition and thesis of the three-dimensionality.

A suitable cutting up of bodies leads to pairs of surfaces which fit each other. Among these surfaces, the *smooth* surfaces have a special position, for which the fit also is maintained during a shift - except for overlappings. Representative of the smooth surfaces, aside from the plane and the sphere, are the cylinder and regular corrugated tin. If any given shift is demanded, in which a pair of contact points on the two surfaces neither remains fixed (rotation symmetrical surfaces), nor is excluded, then only the plane and the sphere are still "smooth" in the sense of the definition.

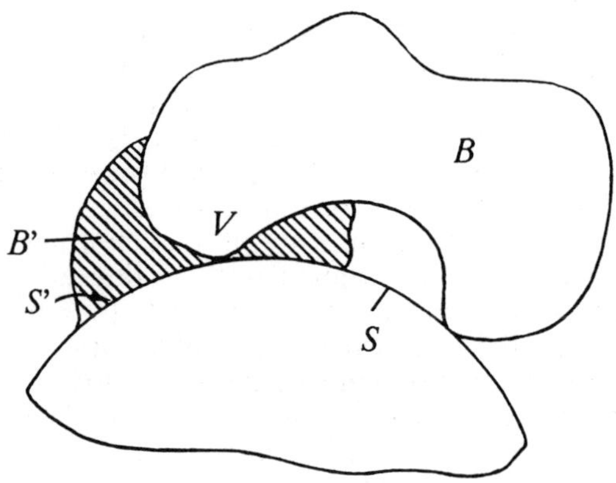

Figure 35

The smooth surfaces already have their own "unequivocality theorem": if a surface is smooth in relation to *one* particular body (solid, but of any given form), then it is smooth in relation to *all* bodies. This proposition is easy to verify: if any given body *B* (Cf. Fig. 35) touches a smooth surface *S* with the vertex *V*, then *B* can be complemented with a body *B'*, which is plastic at first, then rigid, in such a way that *B'* fits the smooth surface *S* and *V* lies inside of the surface *S'* of *B'* which touches the surface *F*. But according to the definition of smooth, *S'* fits *S* at every point, so that *V*, too, can be brought into contact with every point on *S*. With that, it is also true that a point on the given body *B* which does not have a point of contact with the smooth surface, cannot become a contact point during a shift, since if they touched once, they would always touch, as has just been shown. As one can easily imagine, this is only true for the inner point of a smooth surface; the points which lie on the boundary of a smooth surface differ from the inner points in their contact properties. The criterion for shiftability plays a role in relation to cut for the methodical order of the production procedure. Plane cuts were only defined above by the control procedure, but the generation of plane cuts without further aids was left open. This is different, however, with the so-called "*Dreiplatten-Schleifverfahren*" (three-plate grinding procedure), which goes back historically to William K. Clifford and Hugo Dingler.[56] Two simple bodies are considered: if these are ground down, without distinguishing a motion direction, until parts of their outer surfaces fit each other, then one obtains, by definition, smooth surfaces. These will, as a rule, be spherical pieces of surface. What we already know about these bodies is that, whether concave or convex, they prove to be smooth in relation to any given body. Thus, they prove to be smooth *a fortiori*, even in relation to other smooth surfaces. Two (convex) spherical surfaces can only have a single inner pair of contact points. The only inner contact points that two concave surfaces can have form the boundary of the smooth surface on the counterpart. As such they will not be considered here. If, however, a concave surface meets a convex one, the contact possibilities depend upon which of the two has a stronger curvature.

The words "concave" and "convex" have not been defined up to now. Before we rectify this, first let us continue with the production

---

56 All previous literature has considered Hugo Dingler to be the founder of the *Dreiplatten-Schleifverfahren* for the theoretical discussion - which means he discovered that the *Dreiplattenverfahren* makes a circle-free definition of "plane" possible (it is undisputed that the procedure had already been practically applied). Lately, however, it seems to me that William K. Clifford should be given the credit for this. In his book *The Common Sense of the Exact Sciences* (1885), he writes: "The plane surface may be defined as one which is of the same shape all over and on both sides. This property of it is illustrated by the method which is practically used to make such a surface. The method is to take three surfaces and grind them down until any two will fit one another all over." (p. 66)
He adds a discussion of the property of pairwise fitting in shiftability, and the definition of the straight line as the cut-edge of two planes on the same body.

procedure. If one carries out the grinding process on all three plates "simultaneously," i.e., grinds them alternately so that they fit one another pairwise, then two convex surfaces will, when meeting, grind down the inner points, and when coming against a concave surface of greater curvature, will always grind down the rim. As a result, one obtains smooth surfaces which fulfill the definition of plane: each of the three bodies has each of the other two as fitting pieces, which, in turn, also fit each other. The "three-plate grinding procedure" is thus a possibility for a methodically-primary performable production, i.e., production without tools and with planes which have already been realized.

Concave and convex are easily defined with the help of the fitting pieces. A smooth surface is called *concave* when two of its fitting pieces can only touch each other at *one* inner pair of contact points. A smooth surface is called *convex* when its two fitting pieces can only touch each other in such a way that at least one boundary point is contained in every pair of contact points (Cf. Fig. 36). Finally, a smooth surface is called *flat* when it is neither concave nor convex. Obviously, the smooth surfaces which are flat are the plane surfaces.

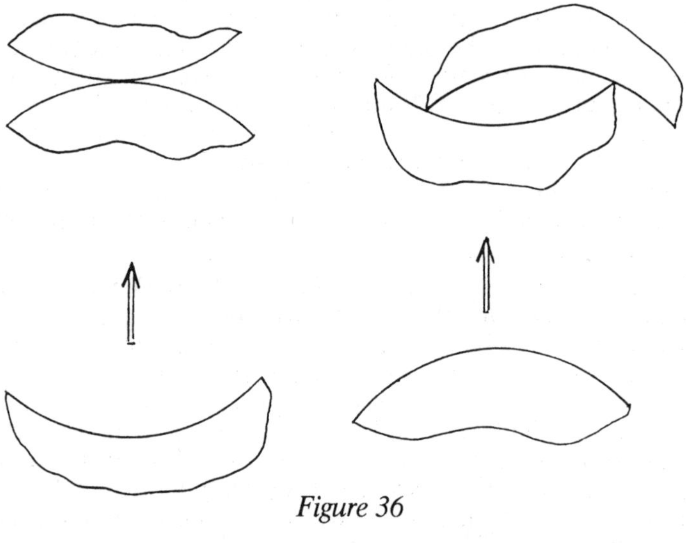

Figure 36

Now all of the means have been brought together which are necessary in order, on one hand, to grasp the connection between the definition of plane and a methodically primary production of plane surfaces, and on the other hand, to show their universal unequivocality. To this end, it must be shown that plane surfaces are smooth. (The reverse conclusion, that smooth surfaces which are also flat are plane surfaces, is asserted above, and is trivially true: when two smooth surfaces which are neither concave nor convex are, as required, brought into contact - hence, exhibit neither exactly one inner pair of contact points, nor none at all - then they must touch each other at all points (except for overlapping). And this is true for each two of the three bodies.)

This conclusion from "plane" to "smooth" cannot be drawn without a certain effort. Since this proof has already been published elsewhere,[57] it shall not be repeated here. The decisive thing here is that plane surfaces as defined (form equality as replaceability with the assistance of a third fitting piece) implies that the fitting - except for possible overlappings - is maintained during a shift. But this means that a partial area of a plane surface which fits its opposite in the initial position, likewise fits its new opposite after a shift - and this is precisely the definition of "smooth."

Since the truth of the special unequivocality theorem for "smooth" has already been demonstrated above, and since all planes are smooth, plane surfaces which have originated in various production contexts can only touch each other in such a way that they both must be regarded as smooth. The only other possibility, which can be ignored, is that planes with different production histories might prove to be concave or convex. This is, however, already logically excluded by the definition of plane.

With that, it has been demonstrated that planes from any given production procedure fit each other - except for overlappings. The form of the plane is universally unequivocal. This is knowledge about the production procedure for planes. The grinding process brings about the property of being smooth, in the sense of the indistinguishability of all points in contact with any given bodies.

The universal unequivocality is also true for right edges and right corners. It can be stated that any two right edges from any given production procedures are equal in form, and likewise for right corners. It is easy to demonstrate that this unequivocality obtains.

According to definition, right edges are formed by two plane surfaces. By exploiting the universal unequivocality of the plane, it has already been proven that the surfaces which form a right edge are equal in form. Thus, it only remains to be shown that right edges from various production procedures have the same angle of opening. Assuming that one has two triplets of right edges which are equal in form: *a*, *b*, *c* and *k*, *l*, *m*. Any two of the three edges *a*, *b*, *c* or *k*, *l*, *m* supplement each other to form a plane. Now assuming that a wedge of the first triplet, say *a*, were "more obtuse" than a wedge of the

57 Cf. P. Janich (1976), p. 116f.

second triplet, say *k*. Then all three wedges of the first triplet would be more obtuse than the respective wedges of the second triplet.

If one now puts two wedges of different triplets together, say *a* and *k*, the result is a form which is "more obtuse" than a plane, and thus does not fit the plane. If one replaces *k* with *b* in this form, that is, a more acute wedge with a more obtuse wedge, then the new form, consisting of *a* and *b*, could only become even more obtuse, and thus not fit the plane. But this contradicts the requirement that *a*, *b* and *c* be right wedges. But if *a*, *b* and *c* cannot, therefore, be more obtuse than *k*, *l*, *m*, then *k*, *l*, *m* cannot be more acute than *a*, *b* and *c*. All six wedges must have the same degree of opening, all right edges are equal in form.

Since right corners are formed from the meeting of three right edges, the form equality of all right corners, independent of their origin in certain production procedures, is already shown. Aside from the plane, the right edge and the right corner are also universally unequivocal.

It follows trivially from this, that any two plane surfaces, any four right edges and any eight right corners supplement each other to form a "full body." That means that none of these form-building surfaces - plane surfaces, right edges and right corners - remains without a fitting counterpart when suitably joined together (with the exception of possible overlappings).

As a result, the prototype-free reproducibility of fundamental spatial forms, which was employed in the proof of the basic theorem of the dimension theory, has been demonstrated. All surfaces, right edges and corners may be treated equally with reference to their cutting possibilities.

It is of epistemological significance that the fitting completibility of plane surfaces, right edges and right corners is *knowledge about the production procedure* of these fundamental spatial forms. It is neither a "natural law" nor a "natural" property of bodies to behave in this way; the described phenomena are first achieved by the technical or artificial generation of the fundamental spatial forms. With that, we can now turn to the question concerning why the layman and the specialist, the engineer and the physicist, are so intuitively certain about the three-dimensionality of the corporeal world; that the three-dimensionality of space is always understood as being, of necessity, true.

THE INTUITIVE CERTAINTY OF THE LAYMAN AND THE SCIENTIST

Examples were given in the foreword to this book which show the intuitive certainty of the layman in reference to the three-dimensionality of space: he distinguishes length, breadth and height on a box, or names orientations in relation to his own body according to the three directions front/back, above/below and left/right. Are these merely language habits which we find in our ordinary language, or can reasons for this speech practice be given from the theory of three-dimensionality which has been developed here?

The psychology of language aspect, that accustomed language habits and the resulting ignorance of alternatives can already produce a diffuse impression of "necessity," shall not be further pursued here. Evidence which is generated through practice in, and familiarity with, a handed-down speech practice, can immediately break down when this speech practice is changed. Historical examples for occasions of such breakdowns span the spectrum: religions and revolutions, new technologies or knowledge, even fashion. The mere fact of normal language usage, the lack of knowledge about alternatives, and psychologically describable motivations of holding something to be true because it is familiar, are not seen here as the "necessity" with which the layman is convinced of the three-dimensionality of space.

The reason for the conviction of the layman that "the space" of the things of daily life is three-dimensional, is based more upon an intuitively correct understanding of the fact that, in this, it is a matter of a form-theoretical state of affairs of a normative character. In order to substantiate this, the everyday meaning of a technical-practical interaction with formed bodies is emphasized. Already small children acquire manual dexterity and a feeling of balance by playing with building blocks which, in the simplest case, are cuboids, cubes and cylinders. The fitting of all blocks which bear plane surfaces, as well as right edges and corners, forms very quickly the basic insight that a practiced child demonstrates. And in the technical, civilized world, a person will never be set free from this practice of dealing with artificially formed bodies, with instruments, furniture, houses, i.e., with artifacts in the broadest sense. Whoever buys a cooking pot for an electrical hot plate expects that the two will fit each other. Whoever acquires a chair or a table expects that these will not wobble, even if they have more than three legs; the number of objects of everyday usage which fit each other or can be stacked is immensely large, and the technical connections are manifold. The plane surface and the right angle are so omnipresent that their repudiation in the construction and furnishing of one's own anthropological philosophy of life is praised as being spiritually becoming.

The technical, practical interaction with bodies on which fundamental geometrical forms are realized, occurs language-free to an essential degree. It has become part of the skills and familiarities of everyday life. Only when *de facto* (and as such neither reflected upon nor reasoned out) existing *expectations are not met*, does the silent assessment of the layman show itself. If the table wobbles on the floor, or if the cooking pot does not fit on the hot plate, this is seen as a lack of quality. The table or the pot (if not the stove or the floor) are then "badly made," i.e., the layman sees the lack of fit as a *defect*, which is to be traced back to the goal of the production being missed because of inattention to the rules of manufacture. The fitting of spatial forms, which is daily taken for granted, proves to be intuitively correctly understood by the way in which a layman deals with faults, namely in its character as the result of production, which can be expected based on standardized actions.

When speech comes into play, the layman speaks about form and magnitude, about position and motion possibilities of bodies and of hollow spaces. This speech corresponds to a technically mastered, non-linguistic practice. This practice, however, is not based on natural properties of bodies or on natural laws, but attaches itself to the successful production of spatial forms. The actual interaction - language-free as well as spoken - with disturbing deviations is a reliable indication for the fact that the layman, beyond any philosophical reflections, has an intuitively correct access to the normative character of spatial forms on bodies, as well as their character as the results of actions.

The "necessity" can then be explained from this without difficulty. It consists first of all of the actual recognition and technical worthiness of the application context for fundamental spatial forms. At all stages of technological development, the use of instruments in which the fitting character of the plane surface and the right angle is not employed is hardly conceivable. Thus, this is a general goal in the technical production of instruments. In other words, the "*necessity*" of three-dimensionality, in the sense of a *non-revisability through any experience*, lies in the normative character of the definition of fundamental geometrical forms. Such definitions are not merely verbal tricks which could also turn out quite differently, but, in the end, rules for the production of spatial forms whose deliberately generated properties are intended to be technically applied. Rules of this kind have the character of principles. They do not assert anything, but rather prescribe something - and rules can also reasonably provide orientation in situations in which they cannot be put into action. As long as the production purposes for fundamental spatial forms are not given up - and not even the greatest technology sceptics demand this - then the three-dimensionality of the body and of hollow space, relative to this purpose, remain, in the sense of the theory presented here, as an immutable, a "necessary," matter of fact.

Not only the layman, but also the engineer and the scientist, here, particularly the physicist and the astronomer, are intuitively certain about the three-dimensionality of space. There is, in fact, a discussion in theoretical physics, in which "dimension" has experienced a manifold expansion of its meaning to include practically every physical perimeter which satisfies certain mathematical criteria. Unimpressed by these possibilities for an expanded meaning, however, engineers, like physicists, use a three-dimensional coordinate system for the localization of bodies and events in space. It does not matter whether it is a Cartesian, a polar, or any other coordinate system. Even when the technical-practical foundation of measuring physics (and its sub-disciplines, like astronomy) is not anchored in the self-conception of the scientists, there is still enough "operational common sense" available to be able to recognize that for bodies which move freely, i.e., those which are neither on a surface nor led along a rail, three spatial coordinates are needed in order to unequivocally describe their position.

It is a misunderstanding of relativity physics worthy of a layman, when one occasionally hears or reads that the revision of the fundamental concepts of space and time of classical physics by Einstein's relativity physics means a transition from a three-dimensional to a four-dimensional space. Einstein himself pointed out again and again that making spatial dimensions relative to the motion situation of the observer and the judgement of spatial distance by the speed of light signals does, in fact, dispense with the temporal independence of the spatial concepts of pure geometry, but does not raise the time perimeter to a fourth dimension which would be equal to the three spatial extensions. And all proposals of the special theory of relativity to define purely spatial distances in a four-dimensional space-time would truly be without any meaning, if the purely spatial ("in place") states of affairs were not unequivocal and describable with the help of geometry, in the first place.

The question (by no means trivial) about how an astronomer on the earth can measure the distance between two stars, and the solution to this problem which is proposed here, offer no occasion to assume that physics or astronomy have strayed from a three-dimensional coordinate system as a foundation for the localization of bodies "in space," *de facto* then always for the localization of bodies in relation to each other. Does the engineer, the physicist or the astronomer take a different foundation for this than the one described above for the layman?

A negation to this question can be verified with the following reasons. First of all, the engineer or the scientist is not born into his profession, but has a long individual learning history to get through before he can dedicate himself to training to become a specialist. Hence, he needs to have already learned speech and action, at least in the perimeters which are required for scientific training. He also requires the elementary spatial distinctions, a knowledge of the fundamentals of geometry, and at least an elementary manual dexterity. In short, every engineer and scientist is also a layman and, as such, has the layman's capacities for action and knowledge concerning the fitting of plane surfaces and right angles, and the localization and volume estimation of bodies. Furthermore, he needs and uses these abilities.

Even the most highly-developed technology and theoretically fine-spun science never cause him to have to put the pragmatic worth of this practice of spatial speaking and acting into question. Whether a house, a telescope, an electrical motor or a computer are to be constructed and built, geometrical words with operationally clear meanings have always been used. The sciences do not, as a rule, reflect this conceptual and practical foundation, although it is precisely reflection about such silent assumptions which has caused the most important radical changes in the history of physics. A difference between laymen and specialists is, at best, the degree of terminological explication and the application of mathematics. Whereas a more or less crude technical state of affairs and its description suffices for the layman, depending on the situation, the specialist often pursues the working out of pretty theories as an end in itself, as well as the refining of production and measurement tolerances. But it would never occur to him that bodies or hollow spaces might have more than three dimensions, or that the position or motion of bodies could be specified in more than three dimensions.

The intuitive certainty of the engineer and the specialist is no different from that of the layman. The "necessity" of the three-dimensionality of empirical space is also not reflected upon by engineers and specialists. It has proven itself. One knows of no alternatives; technical practice would pull the rug out from under any hair-splitting theoretical objection from the very beginning.

The three-dimensionality of space is a nice example of how little the self-conception of the sciences, which today is predominantly an empirical one, holds to the facts of their own practice. Even someone who is convinced that he knows everything that he scientifically knows about the world from experience, will not allow the most important conditions of experience - the three-dimensional localization and description of bodies - to be subject to revision in light of new experiences. There is something like a *practical a priorism*, which is exhibited when one refuses to even allow parts of one's own practice of research in natural science to be made the subject of a critical discussion. To ignore them as a problem is the same as an admission for the empiricists that they do not hold an empirical revision to be possible. Three-dimensionality is true for scientists with practical and principle necessity - no different than for the layman.

The definition and the thesis presented in this book concerning the three-dimensionality of space, were developed from the beginning with the intention of not being restricted to an audience of mathematicians and physicists. Rather, it should develop an understanding of three-dimensionality which does justice to the conviction of the layman and the unreflected parts of the scientist's conviction. What is emphasized again and again in literature about the problem of three-dimensionality as being especially irritating is precisely the fixedness of the conviction that empirical space is three-dimensional, in glaring contrast to the epistemological lack of good reasons for this fixedness.

The answer presented here now takes into account the fact that even the layman does not choose his examples for the three-dimensionality of space from the realm of forms which are indeterminate and unknown in their spatial form, such as clouds, portions of flowing water, or the Northern Lights. It is rather the case that the words length, breadth and height, as well as the directions in reference to the body axes, even the *straight lines* themselves, always are in reference to *plane surfaces* and *right angles*. The three-dimensionality of space, and the irritation caused by the question concerning whether space could not have two or four dimensions (and why), are therefore states of affairs which are bound to distinguished fundamental geometrical forms. The "necessary" and, at least practical, a priori character of three-dimensionality is connected to the fact that for every spatial experience in the world, an order scheme with the fundamental concepts of Euclidean geometry is applied with technical success. Not even the favor that non-Euclidean geometry has found in parts of modern physics has historically changed this state of affairs, nor could anything be changed by the (of itself justified) systematic application of non-Euclidean geometry. The three-dimensionality of space is the cultural accomplishment of raising the world to a higher level of cognitivity through the discovery and technical production of planes and right angles.

THE EPISTEMOLOGICAL STATUS OF THREE-DIMENSIONALITY: DRAWING A BALANCE

In Part I, the most important types of attempts to explain three-dimensionality were discussed, and a new systematic solution proposal was presented in Part II. Now a review of sorts shall answer the question about the epistemological status of the three-dimensionality of empirical space, and which consequences result from this (e.g., for the understanding of geometry). The expression "epistemological status" refers to the reasons from which the truth of three-dimensionality is derived.

Since the clarification of these two questions is, at the same time, a balance of the findings of this book, the core of the historical and systematic results are summarized here:

1. What is meant by the term "three-dimensionality of empirical space"? "Empirical space" is not an individual object for itself, which could be researched by a special science which is qualified for it, or about which something could be known from mere experience. "Empirical space" is rather a concept formed for abbreviated speech about spatial states of affairs. Spatial states of affairs are represented in statements in which "spatial" words appear. Whoever would like to acquire an overview of the set of the possible spatial states of affairs, can do this by writing down a (perfectly finite) list of all the spatial words in his own language.

In this book such a list is not presented. It suffices to name the purpose of speech about spatial states of affairs. And this purpose is to make unambiguous statements possible, about the shape, the magnitude and the position (and therefore the motion possibilities) of bodies and hollow spaces. All the words of ordinary language, as well as mathematical language and that of the natural sciences, which serve this purpose, belong in the list of "spatial" words.

Disregarding the literal meaning of "dimension," which indicates measurement, the three-dimensionality of bodies and hollow spaces was not determined with magnitude theory, but with form theory. Every body can be enclosed in a cuboid in such a way that it touches all six surfaces of the cuboid.

For the definition of cuboid only the words "plane" and "orthogonal" (as a special cut of two planes) are necessary. These fundamental geometrical terms only name spatial forms, and have an unequivocal meaning independent of the magnitude of bodies on which they appear.

The distinction of the triad resulted from the proposition that only three planes can intersect pairwise orthogonally, as long as the "plane" is a form produced on a body, and not merely a formal expression of an "uninterpreted geometry," empty of meaning. The distinction of the triad is based on the insight that (as opposed to magnitude) the plane and the right angle between planes can be technically reproduced at any time, unequivocally and without reference to a distinguished standard, i.e., prototype-free. This is purely prescriptive knowledge which, in the form of a "recipe," provides a normative system of directions for the act of working bodies.

The three-dimensionality of empirical space is thus a result of the effort to make available at any time an unequivocal conceptual and technical possibility for describing the form, the magnitude and the position of bodies.

2. The experience which goes into the thesis of the three-dimensionality of empirical space is not one of natural science, but rather a historical one concerning the success of technical means for the technical purposes of technically shaping the world of bodies. The historical basis of our modern view of three-dimensionality is the cultural-historical accomplishment of having thoroughly established the spatial forms of the plane and the right angle for technical-practical reasons. A cultural history of these fundamental spatial forms must be written as a history of reasons which lays out which technical goals have been served by the manual production of, respectively, the plane and the right angle. This history begins most likely in the area of construction technology and the tools which were required in it. Candidates which might be even older could at best be weapons for hunting, more precisely, the production of "straight" arrows and spears.

The production of smooth and spherical surfaces, of floors, walls and doors, and of every tool, must have been carried out completely theory-free. The fitting of plane surfaces to each other, the completion of right angles between pairs of straight lines to a full line, and the completion of pairs of planes at right edges to full planes, are elementary states of affairs in this, manually mastered quite early, which could simultaneously serve the production practice as a criterion for finished goods. If a barrel on the floor, or the cover of the barrel, wobbles, or if a rectangular clay brick cannot be stacked into a wall in such a way that it fills in its space, this must be seen from the beginning as a flaw in production, and with that, as a failure in the technical purpose of production.

Thus, a pronounced experience in the course of cultural history, informs us that the forms of the plane and the right angle, which are artificially produced on bodies, are suitable for all kinds of technical purposes. With the advancing technical possibilities, these technical purposes have, for their part, also undergone a history of differentiation. The gains in the realm of the technical mastery of spatial forms were cumulative, which deviates from the predominating opinion in the philosophy of the natural sciences today that every gain in knowledge must occur in revolutionary leaps, because they are accompanied by theory. There is no known example which would verify the loss of the producibility of spatial bodily forms through newer, younger technologies.

For the clarification of the epistemological status of three-dimensionality, it is decisive that the knowledge which is needed to this end consists of a technical capacity for action, namely, the production of planes and right angles. This capacity for action attained its full development independent of the theories of natural science. And finally, recipes, like the instructions of a master craftsman to an apprentice, provided the quality and success criteria for the technical practice of the production of spatial forms.

If it is stressed, in an over-emphasis of the empirical position, that a person must first know "from experience" that bodies even exist, that they are movable and workable, and finally that plane surfaces and right angles can be produced on them, this need not be dismissed. Such trivial experiences, which are as indispensable for the learning history of any modern person, as well as for the possibility of natural sciences, are, at any rate, gained theory-free through the practical performance of actions. This means there is no need for some proof, derived from a theory, that a body can be worked in the way mentioned; the practical performance of actions itself insures the necessary trivial experiences with artificially worked bodies.

3. Since the practical acceptance and technical success of the production of instruments cannot be doubted, it must be asked in what sense the thesis of three-dimensionality, which is built upon the technical formation of bodies, possesses "necessity." It obviously cannot be deposed by any new tricks of definition, by any new experiences in natural sciences, or by other kinds of experience. The intuitive certainty of the layman and the natural scientist about being able to unequivocally determine the form, magnitude and position of bodies with three spatial specifications can be seen as legitimate through retrospective philosophical reflection.

The plane, and the right angle between planes, are the two spatial forms which can be produced on bodies, at any time, new and equal, in an exactly specifiable sense. They are the only unequivocal fundamental spatial forms upon which a geometry can be constructed. They are methodically primary, i.e., for their generation, there are known procedures in which no other spatial forms, like the straight line or the sphere, are technical or conceptual prerequisites. (The sphere can, incidentally, also be methodically primary produced, namely by the grinding of two bodies until they mutually fit in one area of their surface. This production is, however, not unequivocal because it is not compelling that the same radius of curvature always results during repetition.)

Seen epistemologically, the three-dimensionality of empirical space is a side effect of a technical and linguistic practice, whose factual universality has a specific logical reason. As theory-free knowledge of the world, only trivial experiences regarding the workability of bodies are needed which are, by virtue of the mere performance of an action, free of doubt. But it can be theoretically recognized - not geometrically or in natural science, but epistemologically - that the production procedures for plane surfaces and right angles allow one to expect the equality of the results. All plane body surfaces and right angles fit each other in the appropriate way, independent of their individual history of production. This property of planes and right angles, called "unequivocality" here, provides the distinction and preference over all alternatives, and assures them the status of "necessity" - in the sense of a non-availability through linguistic conventions and a non-revisability through experiences.

4. The theory of three-dimensionality presented here has consequences for the conception of geometry. First of all, recall the findings which have been made in passing:

Euclid's geometry, indeed, the ancient and modern (synthetic) geometries are built upon the concept of the point. This is the actual, undefined fundamental concept of geometry. The definitions of ancient geometry follow from it, as well as the axioms in modern geometry for lines and surfaces, straight lines and planes. Geometry is first done as planimetry, i.e., geometry of the plane, and only after that, as stereometry. The non-planar forms like sphere, cone, cylinder and the Platonic bodies are formed from, or built upon, plane figures.

The distinction of the point as fundamental concept is carried over into the development of topology as an underlying mathematical subdiscipline for geometry. Topology considers spaces as sets of points and defines the dimension number of such points as invariant in mapping, in the sense of continuous distortions. Stated intuitively, linear figures like a straight line segment or a circle can only be distorted into, again, linear forms like, respectively, an "*S*" or a square, but not into a surface or spatial form. A circular surface can be transformed into a square surface through deformation, but not into a sphere, and so on. Topology, which is built upon the fundamental concept of the point, and the concept of point-spaces which is derived from it, can be clearly characterized by the fact that the shape of the form of a particular dimension number does not matter.

It is self-evident that the theories of geometry and topology which have been developed by mathematicians are free of contradiction, and therefore do not provide any occasion for criticism from a logical standpoint, or from one of modern mathematics. But this modern conception of geometry and its topological foundations has been bought at a high price. First of all, the definition of all the concepts of the theory must consciously be dispensed with. "Point," as fundamental concept, remains undefined (as do a few others, like "set" and "function," by the way.) This abstinence is primarily just a historical fact, but today it is considered by mathematicians to be programmatically dictated, and is occasionally (and incorrectly) justified with the assertion that it is impossible to explicitly define all the concepts of a theory. This assertion can, however, only be verified for a very particular type of definition, namely, for the replacement or abbreviation type of definition where words must occur in the defining portion of the definition which have already been defined. In this way, indeed, no start for the definition of a "fundamental concept" can be gained.

Aside from a lack of conceptual precision, this rejection of definition also sacrifices the ability to answer the question regarding the exact nature of geometry as a science. Besides this, the problem of the application of geometry to the objects of experience remains unclarified.

A finding has been mentioned in passing, that the modern conception of geometry distinguishes between a "mathematical" geometry (in which the relation of fundamental concepts like point, line and surface to the physical world is not an issue) and an "empirical" geometry (in which the choice of a certain geometry from competing alternatives is a question of empirical expediency). The epistemological status of geometrical axioms and propositions noticeably blurs here. Mathematical geometry is considered to be purely conventional, i.e., established through the mere stipulation of axioms and definitions. Only logically contradictory stipulations are "forbidden" in this. Applied geometry is seen, in diffuse ways, as being dependent upon experience. Among these diffuse perspectives are expansions of the concept of experience to operational definitions, as well as the radical-empirical view that the validity of a certain geometry can be decided by measurements and experiments, or the liberal-empirical view that certain geometries have proven themselves historically in the empirical sciences.

The costs of this epistemological imprecision are manifold. Mathematicians cannot justify why they do precisely the geometry which fits the needs of the engineer and the physicist astonishingly well, and not others, which are likewise contradiction-free. And physicists cannot account for how they hope to gain empirical propositions about space, with measuring instruments whose correct functioning can only be described, again, in geometrical propositions. In other words, why is it not disturbing that the spatial states of affairs of measuring instruments - also objects of experience - must already be fixed before the empirical determination of spatial states of affairs can begin?

In Chapter Six, this was spoken of as "Euclid's heritage." Euclid's heritage is the construction of geometry from the point to spatial forms of higher dimensions. The objects of geometry originate in the context of the act of drawing, as is indicated by the Greek words for point and line. The fact that, to this end, a drawing plane, a straight ruler and a compass - hence, bodies with certain spatial properties - are necessary, did not play a role for Euclid. The slave activity of craftsmanship, and the properties of the instruments which were due to the handworker, and which are necessary for a drawing planimetry, could not play a role for the ancient academicians. The aspect of technical practice is lacking in ancient geometry, as in the ancient natural science of Aristotle, for instance, and it first gains significance in the measuring and experimental physics of the 17th Century with Galileo. But while modern physics as a laboratory science has long overcome any contempt for technical-practical things, Euclid's academic heritage continues to have an effect in mathematics. The constitution of the objects of geometry through the production of drawing tools does not play any role.

This deficit shows itself even more clearly in the realm of topology, in which the metaphor of the cut through spaces as subsets of the cut space is quite common, but in which a reference to the elementary human activities of cutting or dividing is never made. If this would happen, as it does in this book, then the definitional constitution of the objects of topology and geometry would begin with the bodies of our world of daily experience. Hence, if mathematics were to get involved with the epistemological status of the objects of topology and geometry, it would only make sense to choose the forms of the plane and the right angle as the "most simple" or "fundamental concepts." The consideration of a physical body as a set of points, in contrast, would more likely be erroneous in this perspective.

Euclid's heritage gives preference to the point as the fundamental concept and, viewing geometry as a theory of sets of points, provides precisely the occasion for disregarding the spatial form. By overcoming Euclid's heritage - by constructing geometry in a descending sequence from bodies, through surfaces and lines, to points - the high costs mentioned here can be avoided, without the loss of mathematical insights and theories, and without dispensing with recognized golden rules of the mathematician guild.

The expression "Euclid's heritage" has furthermore proven to be ambiguous. Until now we have spoken about what Euclid has bequeathed the history of mathematics and ideas through his ascending construction of geometry from the point to the body. Historical justice, however, demands that Euclid himself be seen as a heritage. The triad of boundaries (point, line, and surface as the boundaries of, respectively, the line, surface, and body) and the triad of the bounded (line, surface, and body) was not invented by Euclid or Eudoxos. The authors of these distinctions are unknown. The answers to the questions about the reasons and ways in which the words for surfaces, edges and corners, for round and straight lines, for obtuse, right and acute angles, came into use, lie in the darkness of the early history of language, which is even difficult to illuminate with the aid of etymological research. Euclid, in any case, discovered them already in use and, restricted to the practice of drawing, applied them in his geometry. The historical roots of the triad of geometrical objects are, thus, unknown. It is connected with the development of production and the use of tools. It could therefore assert itself unendangered throughout cultural history, even if the sciences like mathematics no longer perceive its foundation and, especially in form-invariant topology, have made a relationship of the ideal geometrical concepts point, line and surface to the corners, edges and surfaces of worked bodies impossible. Thus, "Euclid's heritage" as the legacy for modern mathematics is only the organization of geometrical theory, and is itself due to an uncomprehended legacy which was left behind by the creators of our language for spatial states of affairs.

5. The theory of three-dimensionality presented here does not only have consequences for the conception of geometry, but also for the distinction of justifications of truth within epistemology. An adequate insight into the nature of three-dimensionality was not only prevented by Euclid's heritage, but, considered from the history of effects, primarily by Kant's heritage as well.

It is due to Kant that two pairs of opposites have determined the division of justification of truth clear up to today, namely the opposites a priori/a posteriori and analytic/synthetic. The first pair of opposites names the distinction that a statement is either true "before all experience" i.e., independent of experience and making experience possible, or is based on experience. The latter names the distinction between a statement which already is true based on the literal meaning of its parts (Example: This bachelor is unmarried.) or a statement which asserts something new which does not follow from the literal meaning of its words (Example: This bachelor is 38 years old.). Analytical sentences are always a priori, since the history of learning the meaning of words is not what is meant by "experience." All empirical propositions, thus, all a posteriori statements, are synthetic for the same reason. The third combination (decisive for Kant's philosophy) from the two pairs of opposites is the synthetic/a priori truth. According to Kant, it is precisely geometry which provides the most important example of statements which are synthetically true a priori.

Kant's theory of the synthetical a priori - as a consequence of relativity physics and its reinterpretation of the foundations of space and time - has led to an overreaction in the philosophy of science of this century, more precisely in its main trends of Logical Empiricism and Analytical Philosophy. "Logical Empiricism" only wanted to recognize statements as scientific which are either logically or empirically true, hence, in Kant's sense, are either analytical or a posteriori. As lamented above, this led to the splitting up of geometry into a mathematical geometry which was only logically true, and an empirical geometry which was only based on experience. Analytical Philosophy, with its representatives like Willard Van Orman Quine, as a counter-trend, lapsed into the opposite extreme, and even raised doubts about the sharp division between analytical and a posteriori truth.

Hence, Kant's heritage, added to Euclid's heritage, and including its exaggerated counter-reactions, thoroughly obstructed an epistemological comprehension of the foundations of geometry and, therefore, of three-dimensionality as well.

The assertion, which has been made often in this book, that neither the layman nor the scientist is prepared to view the three-dimensionality of empirical space as merely definitionally conventional or as merely empirically available, is, at first, also a tribute to the dogma of Logical Empiricism. Unfortunately, the third Kantian path between the two extremes of calling three-dimensionality a synthetic truth a priori, cannot be chosen as a way out here. True, it was attempted above to explain the "necessity" of three-dimensionality, and necessary truth is supposed to distinguish a priori truth according to Kant. But what we have here is more a harmony of words than a equivalence of meaning.

While Kant reflected over the subjective conditions of knowledge in a mentalistic or psychological sense, in the theory of three-dimensionality presented here, it is rather a matter of a public technical practice, led by linguistic communication under historical conditions. The awareness of the linguistic conditions of the possibility of truth of geometrical propositions was first created by the philosophy of language of Frege, Russell and Wittgenstein. The pragmatic deficit of Kant's philosophy (which remained loyal to Euclid's heritage and did not see the technical generation of the objects of geometry) could only be lifted after Husserl's Phenomenology, Dingler's Methodical Pragmatism, and modern theories of action.

As an end result, the theory of three-dimensionality presented here cannot give a general theoretical indication - but it can at least provide a representative individual case - of which intermediate levels of distinction are possible and reasonable between purely analytical statements, which are based only on definition, and purely empirical ones, which are only based on current individual experience (it is, indeed, still reasonable to make such a distinction). The three-dimensionality of empirical space is, in terms of these extremes, neither something purely abstract, nor something found in nature. It is neither a logical law, nor a natural one, nor is it based on one of the two. Three-dimensionality is much more a by-product of human practice, with increasing differentiation directed more and more by theory. Whoever can live in a paradise free of needs, isolated from every technical civilization and every culture with a history of ideas, whoever has no need of tools and instruments, and is essentially inactive, also does not need plane surfaces and right angles on bodies. Only those who actively deal with spatial states of affairs (in technical practice and/or theoretically) are compelled to be involved with the three-dimensionality of empirical space. Thus, whoever pursues such technical and/or theoretical purposes, and is concerned with the success of his actions, and whoever practices a bit of applied geometry to that end, can then convince himself, on the basis of the theory presented here, that empirical space is three-dimensional, and that our intuition regarding this is legitimate.

Appendices

BIBLIOGRAPHY

Alexandroff, P.: "Über den allgemeinen Dimensionsbegriff und seine Beziehungen zur elementaren geometrischen Anschauung", in: *Mathematische Annalen* 98, 1928, 617-636.

Apostle, H. G.: *Aristotle's Philosophy of Mathematics*, Chicago, 1952.

Aristotle: "On the Heavens", in: J. Barnes (ed.), *The Complete Works of Aristotle*, Vol. I, Princeton, 1985, 447-511.

Becker, O.: "Beiträge zur phänomenologischen Begrundung der Geometrie und ihrer physikalischen Anwendungen", in: E. Husserl (ed.), *Jahrbuch für Philosophie und phänomenologische Forschung* VI, 1923, 385-560.

Bolzano, B.: "Versuch einer objektiven Begründung der Lehre von den drei Dimensionen des Raumes", in: *Abhandlungen der Kgl. Böhmischeen Gesellschaft der Wissenschaften*, 5. Folge, Heft 3, Prag, 1845, 201-215.

Brentano, F.: "Zur Lehre von Raum und Zeit", in: *Kant-Studien* 25, 1920, 1-23.

Brotman, H.: "Könnte der Raum vierdimensional sein?", in: E. v. Savigny (ed.), *Philosophie und normale Sprache*, Freiburg/München, 1969, 50-62.

Brouwer, L. E. J.: "Über den natürlichen Dimensionsbegriff", in: H. Freudenthal (ed.), *L. E. J. Brouwer. Collected Works*, Vol. II, Amsterdam, 1976. First appeared in: *Journal für reine und angewandte Mathematik* 142 (1913), 146-152.

Brouwer, L. E. J.: "Beweis der Invarianz der Dimensionszahl", in: H. Freudenthal (ed.), *L. E. J. Brouwer. Collected Works*, Vol. II, Amsterdam, 1976. First appeared in: *Mathemathische Annalen* 70, 1911, 161-165.

Brouwer, L. E. J.: "Bemerkungen zum natürlichen Dimensionsbegriff", in: H. Freudenthal (ed.), *L. E. J. Brouwer. Collected Works*, Vol. II, Amsterdam, 1976.

Büchel, W.: *Philosophische Probleme der Physik*, Freiburg/Basel/Wien, 1965.

Buroker, J. V.: *Space and Incongruence. The Origin of Kant's Idealism*, Dordrecht/Boston/London, 1981.

Carnap, R.: "Dreidimensionalität des Raumes und Kausalität. Eine Untersuchung über den logischen Zusammenhang zweier Fiktionen", in: *Annalen der Philosophie und philosophischen Kritik* 4, 1924, 105-130.

Cleve, J. v.: "Right, Left, and the Fourth Dimension", in: *The Philosophical Review* XCVI, 1987, 33-68.

Clifford, W. K.: *Seeing and Thinking*, London/New York, 1890.

Dingler, H.: *Die Grundlagen der angewandten Geometrie*, Leipzig, 1911.

Dingler, H.: "Über den Zirkel in der empirischen Begründung der Geometrie", in: *Kantstudien* 30, 1925, 310-330.

Dingler, H.: *Die Grundlagen der Geometrie. Ihre Bedeutung für Philosophie, Mathematik, Physik und Technik*, Stuttgart, 1933.

Eddington, A. S.: *Fundamental Theory*, Cambridge, 1953.

Ehrenfest, P.: "Welche Rolle spielt die Dreidimensionalität des Raumes in den Grundsätzen der Physik?", in: *Annalen der Physik* 61, 4. Folge, 1920, 440-446.

Einstein, A.: "Betrachtungen über die Welt als Ganzes", in: E. Wildhagen (ed.), *Materie und Energie. Vom naturwissenschaftlichen Weltbild der Gegenwart*, Berlin, 1932, 338-347.

Fliedner, D.: "Zum Problem des vierdimensionalen Raumes. Eine theoretische Betrachtung aus historisch-geographische Sicht", in: *Philosophia Naturalis* 18, 1980, 388-412.

Galilei, G(alileo): *Dialogue Concerning the Two Chief World Systems - Ptolemaic and Copernican*, (trans.) S. Drake, Berkeley/Los Angeles, 1953.

Gilman, B. I.: "On the nature of dimension", in: *The Journal of Philosophy* 25, 1928, 561-575.

Gray, R.: "Berkeley's Theory of Space", in: *Journal of the History of Philosophy* 16, 1978, 415-434.

Heath, Sir T. L. (ed.): *The thirteen books of Euclid's Elements*, New York, 1956.

Hegel, G. W. F.: *Encyclopädie der philosophischen Wissenschaften im Grundrisse*, Leipzig, 1920.

Hellpach, W.: "Dimensionen in Raum und Zeit", in: *Philosophia Naturalis* 1, 1950, 179-188.

Herbart, J. F.: "Vom körperlichen Raume", in: *J. F. Herbart. Sämtliche Werke*, Vol. XIII, Langensalza, 1893, 151-158.

Horstmann, R. P.: "Raumanschauung und Geometrie", in: *Ratio* 18, 1976, 16-28.

Inhetveen, R.: *Konstruktive Geometrive. Eine formentheoretische Begründung der euklidischen Geometrie*, Mannheim/Wien/Zürich, 1983.

Jammer, M.: *Das Problem des Raumes*, Darmstadt, 1960.

Janich, P.: "Zur Protophysik des Raumes", in: G. Böhme (ed.), *Protophysik*, Frankfurt, 1976, 83-130.

Janich, P.: "Was heißt eine Geometrie operative begründen?", in: W. Diederich (ed.), *Zur Begründung physikalischer Geo- und Chronometrien*, Bielefeld, 1979, 59-77.

Janich, P.: *The Protophysics of Time*, (trans.) R. Brown, Dordrecht, 1985.

Janich, P.: "Natur und Handlung. Über die methodischen Grundlagen naturwissenschaftlicher Erfahrung", in: O. Schwemmer (ed.), *Vernunft, Handlung und Erfahrung*, München, 1981, 69-84.

Kant, I.: "Gedanken von der wahren Schätzung der lebendigen Kräfte und Beurteilung der Beweise derer sich Herr von Leibniz und andere Mechaniker in dieser Streitsache bedient haben, nebst einigen vorhergehenden Betrachtungen welche die Kraft der Körper überhaupt betreffen", Königsberg, 1746.

Kant, I.: "Von dem ersten Grunde des Unterschiedes der Gegenden im Raume", Königsberg, 1768.

Kant, I.: *Kritik der reinen Vernunft*, Riga, 1787.

Lange, F. A.: *Geschichte des Materialismus und Kritik seiner Bedeutung in der Gegenwart*, Zweites Buch: *Geschichte des Materialismus seit Kant*, Frankfurt, 1974, (first published in 1866).

Leibniz, G. W.: "Opera Philosophica" etc., Teil I, J. E. Erdmann (ed.), Berlin, 1860.

Leibniz, G. W.: *Die Theodicee,* A. Buchenau (ed.), Leipzig, 1925.

Lorenzen, P.: *Elementargeometrie. Das Fundament der Analytischen Geometrie*, Mannheim/Wien/Zürich, 1984.

Lucas, J. R.: "Euclides ab omni naevo vindicatus", in: *British Journal for the Philosophy of Science* 20, 1969, 1-11.

Natorp, P.: *Die logischen Grundlagen der exakten Wissenschaften*, Leipzig/Berlin, 1910.

Pitzker, F.: "Die dreifache Ausdehnung des Raumes", in: *Unterrichtsblätter für Mathematik und Naturwissenschaften* 8, 1902, 39-41.

Poincaré, H.: *The Value of Science*, New York, 1958.

Poincaré, H.: "Pourquoi l'espace a trois dimensions", in: *Revue de Metaphysique et de Morale* 20, 1912, 483-504.

Simplicius: *Simplicii in Aristotelis Physicorum Libros Quattuor Priores Commentaria*, H. Diels (ed.), Berlin, 1882.

Suppes, P.: "Is visual space Euclidian?", in: *Synthese* 35, 1977, 397-421.

Thüring, B.: *Die Gravitation und die philosophischen Grundlagen der Physik*, Berlin, 1967.

Trendelenburg, A.: *Logische Untersuchungen*, Leipzig, 1862.

Uexküll, J. v., and Kriszat, G.: *Streifzüge durch die Umwelten von Tieren und Menschen/Bedeutungslehre*, Frankfurt, 1970.

Vollmer, G.: *Evolutionäre Erkenntnistheorie*, 2nd unabridged ed., Stuttgart, 1980.

Vollmer, G.: *Was können wir wissen?* Vol. I: *Die Natur der Erkenntnis*, Stuttgart, 1985.

Vollmer, G.: *Was können wir wissen?* Vol. II: *Die Erkenntnis der Natur*, Stuttgart, 1986.

Weyl, H.: "Gravitation und Elektrizität", in: *Sitzungsberichte der Preußischen Akademie der Wissenschaften*, Berlin, 1918, 465-480.

Weyl, H.: "Eine Erweiterung der Relativitätstheorie", in: *Annalen der Physik* 59, (4. Folge), 1919, 101-133.

Weyl, H.: *Raum - Zeit - Materie*, Berlin, 1918.

Whitrow, G. J.: "Why physical space has three dimensions", in: *British Journal for the Philosophy of Science* 6, 1955/56, 13-31.

INDEX OF NAMES

SUBJECT INDEX

The University of Western Ontario Series in Philosophy of Science

1. J. Leach, R. Butts and G. Pearce (eds.): *Science, Decision and Value.* 1973
   ISBN 90-277-0239-X; Pb 90-277-0327-2
2. C. A. Hooker (ed.): *Contemporary Research in the Foundations and Philosophy of Quantum Theory.* 1973
   ISBN 90-277-0271-3; Pb 90-277-0338-8
3. J. Bub: *The Interpretation of Quantum Mechanics.* 1974
   ISBN 90-277-0465-1; Pb 90-277-0466-X
4. D. Hockney, W. Harper and B. Freed (eds.): *Contemporary Research in Philosophical Logic and Linguistic Semantics.* 1975
   ISBN 90-277-0511-9; Pb 90-277-0512-7
5. C. A. Hooker (ed.): *The Logico-algebraic Approach to Quantum Mechanics.*
   Vol. I: Historical Evolution. 1975 ISBN 90-277-0567-4; Pb 90-277-0613-1
   Vol. II: Contemporary Consolidation. 1979
   ISBN 90-277-0707-3; Pb 90-277-0709-X
6. W. L. Harper and C. A. Hooker (eds.): *Foundations of Probability Theory, Statistical Inference, and Statistical Theories of Science.*
   Vol. I: Foundations and Philosophy of Epistemic Applications of Probability Theory. 1976 ISBN 90-277-0616-6; Pb 90-277-0617-4
   Vol. II: Foundations and Philosophy of Statistical Inference. 1976
   ISBN 90-277-0618-2; Pb 90-277-0619-0
   Vol. III: Foundations and Philosophy of Statistical Theories in the Physical Sciences. 1976 ISBN 90-277-0620-4; Pb 90-277-0621-2
7. C. A. Hooker (ed.): *Physical Theory as Logico-operational Structure.* 1979
   ISBN 90-277-0711-1
8. J. M. Nicholas (ed.): *Images, Perception, and Knowledge.* 1977
   ISBN 90-277-0782-0
9. R. E. Butts and J. Hintikka (eds.): *Logic, Foundations of Mathematics, and Computability Theory.* Part One: Logic, Foundations of Mathematics, and Computability Theory. 1977 ISBN 90-277-0708-1
10. R. E. Butts and J. Hintikka (eds.): *Logic, Foundations of Mathematics, and Computating Theory.* Part Two: Foundational Problems in the Special Sciences. 1977 ISBN 90-277-0710-3
11. R. E. Butts and J. Hintikka (eds.): *Logic, Foundations of Mathematics, and Computability Theory.* Part Three: Basic Problems in Methodology and Linguistics. 1977 ISBN 90-277-0829-0
12. R. E. Butts and J. Hintikka (eds.): *Logic, Foundations of Mathematics, and Computability Theory.* Part Four: Historical and Philosophical Dimensions of Logic, Methodology and Philosophy of Science. 1977 ISBN 90-277-0831-2
    Set (9-12) ISBN 90-277-0706-5
13. C. A. Hooker J. J. Leach and E. F. McClennen (eds.): *Foundations and Applications of Decision Theory.*
    Vol. I: Theoretical Foundations. 1978 ISBN 90-277-0842-8
    Vol. II: Epistemic and Social Applications. 1978 ISBN 90-277-0844-4

The University of Western Ontario Series in Philosophy of Science

14. R. E. Butts and J. C. Pitt (eds.): *New Perspectives on Galileo*. 1978 ISBN 90-277-0859-2; Pb 90-277-0891-6
15. W. L. Harper, R. Stalmaker and G. Pearce (eds.): *Ifs. Conditionals, Belief, Decision, Chance, and Time*. 1981 ISBN 90-277-1184-4; Pb 90-277-1220-4
16. J. C. Pitt (ed.): *Philosophy in Economics*. 1981 ISBN 90-277-1210-7; Pb 90-277-1242-5
17. M. Ruse: *Is Science Sexist?* And Other Problems in the Biomedical Sciences. 1981 ISBN 90-277-1249-2; Pb 90-277-1250-6
18. N. Rescher: *Leibniz's Metaphysics of Nature*. A Group of Essays. 1981 ISBN 90-277-1252-2; Pb 277-1253-0
19. L. Laudan: *Science and Hypothesis*. Historical Essays on Scientific Methodology. 1981 ISBN 90-277-1315-4
20. W. R. Shea (ed.): *Nature Mathematized*. Historical and Philosophical Case Studies in Classical Modern Natural Philosophy. 1983 ISBN 90-277-1402-9
21. M. Ruse (ed.): *Nature Animated*. Historical and Philosophical Case Studies in Greek Medicine, 19th-Century and Recent Biology, Psychiatry, and Psychoanalysis. 1983 ISBN 90-277-1403-7
22. W. R. Shea (ed.): *Otto Hahn and the Rise of Nuclear Physics*. 1983 ISBN 90-277-1584-X
23. H. F. Cohen: *Quantifying Music*. The Science of Music at the First Stage of the Scientific Revolution (1580-1650). 1984 ISBN 90-277-1637-4
24. R. E. Butts: *Kant and the Double Government Methodology*. Supersensibility and Method in Kant's Philosophy of Science. 1984 ISBN 90-277-1760-5; Pb 90-277-2384-2
25. J. R. Brown (ed.): *Scientific Rationality*. The Sociological Turn. 1984 ISBN 90-277-1812-1
26. F. Wilson: *Explanation, Causation and Deduction*. 1985 ISBN 90-277-1856-3
27. J. C. Pitt (ed.): *Change and Progress in Modern Science*. 1985 ISBN 90-277-1898-9
28. H. B. Hollinger and M. J. Zenzen: *The Nature of Irreversibility*. A Study of Its Dynamics and Physical Origins. 1985 ISBN 90-277-2080-0
29. K. Okruhlik and J. R. Brown (eds.): *The Natural Philosophy of Leibniz*. 1985 ISBN 90-277-2145-9
30. G. Oddie: *Likeness to Truth*. 1986 ISBN 90-277-2238-2
31. F. Wilson: *Laws and Other Worlds*. A Human Account of Laws and Counterfactuals. 1986 ISBN 277-2232-3
32. J. Earman: *A Primer on Determinism*. 1986 ISBN 90-277-2240-4; Pb 90-277-2241-2
33. R. E. Butts (ed.): *Kant's Philosophy of Physical Science*. Metaphysische Anfangsgründe der Naturwissenschaft (1786-1986). 1986 ISBN 90-277-2309-5; Pb 90-277-2310-9

The University of Western Ontario Series in Philosophy of Science

34. I. B. MacNeill and G. J. Umphrey (eds.): *Advances in the Statistical Sciences.* Festschrift in Honor of Professor V.M. Joshi's 70th Birthday (6 volumes). Vol. I: Applied Probability, Stochastic Processes, and Sampling Theory. 1987 ISBN 90-277-2393-1
35. I. B. MacNeill and G. J. Umphrey (eds.): *Advances in the Statistical Sciences.* Vol. II: Foundations of Statistical Inference. 1987 ISBN 90-277-2394-X
36. I. B. MacNeill and G. J. Umphrey (eds.): *Advances in the Statistical Sciences.* Vol. III: Time Series and Econometric Modelling. 1987 ISBN 90-277-2395-8
37. I. B. MacNeill and G. J. Umphrey (eds.): *Advances in the Statistical Sciences.* Vol. IV: Stochastic Hydrology. 1987 ISBN 90-277-2396-6
38. I. B. MacNeill and G. J. Umphrey (eds.): *Advances in the Statistical Sciences.* Vol. V: Biostatistics. 1987 ISBN 90-277-2397-4
39. I. B. MacNeill and G. J. Umphrey (eds.): *Advances in the Statistical Sciences.* Vol. VI: Actuarial Science. 1987 ISBN 90-277-2398-2
    Set (Nos. 34-39) ISBN 90-277-2399-0
40. N. Rescher: *Scientific Realism.* A Critical Reappraisal. 1987 ISBN 90-277-2442-3; Pb 90-277-2528-4
41. B. Skyrms and W. L. Harper (eds.): *Causation, Chance, and Credence.* 1988 ISBN 90-277-2633-7
42. W. L. Harper and B. Skyrms (eds.): *Causation in Decision, Belief Change and Statistics.* 1988 ISBN 90-277-2634-5
43. R. S. Woolhouse (ed.): *Metaphysics and Philosophy of Science in the 17th and 18th Centuries.* Essays in Honor of Gerd Buchdahl. 1988 ISBN 90-277-2743-0
44. R. E. Butts and J. R. Brown (eds.): *Constructivision and Science.* Essays in Recent German Philosophy. 1989 ISBN 0-7923-0251-6
45. A. D. Irvine (ed.): *Physicalism in Mathematics.* 1989 ISBN 0-7923-0513-2
46. J. van Cleve and R. E. Frederick (eds.): *The Philosophy of Right and Left.* Incongruent Counterparts and the Nature of Space. 1991 ISBN 0-7923-0844-1
47. F. Wilson: *Empiricism and Darwin's Science.* 1991 ISBN 0-7923-1019-5
48. G. G. Brittan, Jr. (ed.): *Causality, Method and Modality.* Essays in Honor of Jules Vuillemin. 1991 ISBN 0-7923-1045-4
49. W. Spohn, B. C. van Fraassen and B. Skyrms (eds.): *Existence and Explanation.* Essays in Honor of Karel Lambert. 1991 ISBN 0-7923-1252-X
50. J. C. Pitt: *Galileo, Human Knowledge, and the Book of Nature.* Method Replaces Metaphysics. 1992 ISBN 0-7923-1510-3
51. V. Coelho (ed.): *Music and Science in the Age of Galileo.* 1992 ISBN 0-7923-2028-X
52. P. Janich: *Euclid's Heritage: Is Space Three-Dimensional?* 1992 ISBN 0-7923-2025-5

KLUWER ACADEMIC PUBLISHERS – DORDRECHT / BOSTON / LONDON